THE CODE OF LIFE

THE
CODE
OF
LIFE

ERNEST BOREK, 1911—

Revised Edition

COLUMBIA UNIVERSITY PRESS

NEW YORK AND LONDON

To
Hans Thacher Clarke
eminent chemist, master stylist,
exacting preceptor

PREFACE

In writing a sequence of books on science the burden for the author is not the dearth of ideas for the text, but rather for the preface. Science is ever growing; the motivations for writing about it remain constant. But even a preface can offer a challenge: one can try to clothe old ideas in new sentences.

The motives for writing a book on science for the lay reader are to entertain, to enlighten, and to marshal his aid in the cause of science. The first two goals are obvious. Science is at present the richest and most yielding frontier for the human mind. A sense of excitement and joy can be shared even from its passive contemplation. After all, can we not derive almost as much pleasure from reading a fine poem as from writing one? Moreover, it is not the form but the source of creativity which compels our attention. Be it a song, a poem, a fresh idea expressed in a polished paragraph, or a revelation of the world we live in leaping forth from a well-designed experiment, they all surge, through different spouts, from the same wellspring: the creative human soul. The fruits of a scientist's hand and mind can be savored by anyone of intelligence and sensitivity, anyone who can enjoy a poem.

The debate about "two cultures" is as groundless as the wording of the proposition is imprecise. The dichotomy is not between two cultures; it is between culture and no culture. I am not a late-comer to this discussion. About ten years ago, there was one of these "Let us save the humanities" editorials in the *Saturday Review of Literature*, and it was sufficiently irritating to catalyze a letter to the editor. The following is an excerpt from what was printed:

Those are rather moldy straw men Mr. X exhumed for his editorial, "Are the Humanities Worth Saving." I think one would be hard put to find an outstanding scientist who still worships in Mr. Huxley's monolithic church where Science was enshrined as the culture of the age. Indeed, the scientists of today—Einstein, Pauling, Urey, Oppenheimer, to name but a few—are the foremost champions who are striving for the preservation of the whole spectrum of values created by man.

But one can only say amen to Mr. X's conclusion that it is not very helpful to categorize the Sciences as "humane" only when they are taught as cultural rather than professional subjects. The question is not the ultimate goal of the teaching but the scope of the teaching. A chemistry professor is a mere mechanic if he simply tells his students that Mendeleyev aligned the known elements in order of increasing atomic weight and finding gaps in the sequence predicted that unknown elements which could fill those gaps must exist. However, if the professor goes on to emphasize that Mendeleyev's prediction was a bold affirmation of faith that there is a pattern, a reason in nature, then he is not a mechanic but what Mr. X would call a humanist. On the other hand, if a professor of English spends his energies on the *minutiae* of someone else's writings then he is not a humanist but a mechanic. And I rather think that the incidence of mechanics among scientists is not much greater than among humanists.

If science is not another culture, it is certainly a largely hidden one. That this is so is as much the fault of scientists as of anyone else. Until World War II, science had been practiced in our country as a hobby-like activity by members of an elite club. To maintain the posture of exclusiveness a doctrine of arcane profundity was promulgated: science was declared out of bounds to all but the select few practitioners of it. This, of course, is nonsense spawned by arrogance. As the German philosopher Ludwig Wittgenstein said: "Everything that can be thought at all can be thought clearly. Everything that can be said can be said clearly."

A clue to the source of this arrogance on the part of some may perhaps be found in the maxim of the German philosopher, Nietzche: "He who knows himself to be profound endeavors to be clear; he who would like to appear profound to the crowd endeavors to be obscure."

A posture of exclusiveness is particularly outdated these days when the overwhelming portion of scientific activity is financed by our federal government and therefore by the creators of our total wealth, the American people. Science and its products are as much the property of the American public as is TVA or the Hoover Dam or a submarine propelled by atomic power. Yet, any attempt to inform the public, be it via an intelligently written newspaper article or by a book such as this one, is still greeted by snickers by some practitioners of science. Thus, some of the scientists themselves are the first enemy of science against whom the enlightened public's aid must be enlisted.

The resistance of some people to the tides of which they are a component part is not unique to the workers in the field of science. Leon Blum, who was to become a Premier of France, had a hand in defending Dreyfus against the fascist officer-clique of the time. From intimate knowledge of his hero Blum came to the sad conclusion that Dreyfus the man did not live up to Dreyfus the symbol. He said: "If Dreyfus had not been Dreyfus he would have been an anti-Dreyfusard." So, today, many who earn a livelihood from science are, at heart, antiscience.

With the increasing channeling of research financing through the United States Treasury, a new enemy of science has loomed up. That some of our lawmakers would sooner or later become restive in their passive role and would want to become the controllers not only of funds but of science itself was predictable.

Four years ago, in another book, *The Atoms Within Us*, I wrote:

Men with managerial bent like to have a tidy picture in front of them: a simple goal set, the path to the goal outlined, and the outcome guaranteed. But tidiness is a minor virtue and simple goals are the yearning of pedestrian minds. We would still be trying to convert base metals into gold and cure every ailment with poultices and blood-letting if during the past few centuries the creative imagination of all scientists had been shackled to goals set for them by their masters, whether bishops, kings, or congressmen. Since the federal government underwrites more than half of the scientific research in our country, it is no exaggeration to state that our progress in science and its eventual technological applications depends upon the resolution of the impasse between the managerial mind of the congressman and the creative mind of the scientist.

Since that paragraph was written, enormous pressure has been brought on government agencies which support scientific research to award grants on a geographical (read "pork barrel") basis rather than on the merit of the individual investigator. Also, about a year ago there was an actual instance of a congressman giving an order in public to the National Institutes of Health to study the perspiratory effects of vinegar because that condiment always makes him sweat.

Clearly a dual task of educating the public and our lawmakers is an urgent need. Such education must not be tainted with aristocratic snobbery and have as its aim reaching out to the well-schooled few. An appreciation of the aims, the content, the spirit of science is within the reach of any intelligent individual even with limited formal education. It is well known that one of the staunchest and most understanding friends of science in Washington is a congressman who is a former bricklayer, Mr. Fogarty of Rhode Island.

Finally, the reader may be curious to know why a *particular* scientist chooses to popularize his field. If the reader shares with me a distrust of missionary zealots, I wish to quiet his apprehension. This book is simply the result of a genetic coincidence. I happen to enjoy writing a paragraph which is precise in meaning almost as much as performing a telling experiment. Thus, sharing the excitement of my field with the reader is not a crusading chore, it is an enduring pleasure.

ERNEST BOREK

New York
December, 1964

ACKNOWLEDGMENTS

I AM grateful to friends who graciously shared my labors.

Mr. Robert Tilley of the Columbia University Press read each chapter from the point of view of the lay reader and tried to lead me back to within bounds of readability whenever I strayed into the jargon of scientese. He thus served both as a guinea pig and a guide. Drs. Michael Heidelberger and Rollin Hotchkiss very kindly took time out from their laboratories to reminisce about their former associate, the late Oswald Avery. My colleague, Dr. P. R. Srinivasan, and two former students, Drs. Ann Ryan and Erwin Fleissner, read the manuscript and corrected, it is hoped, most of my errors. Mrs. Irene Marsh and Dr. Geraldine Poppa were, as before, my trusted proofreaders.

Finally, I am grateful to Dr. Clarke, to whom this book is dedicated, for a critical reading of the manuscript for both style and content.

CONTENTS

1 HOMAGE TO THE PIONEERS
OR, OF SWEET PEAS AND PUS CELLS

"LIKE BEGETS LIKE." This truism has been part of man's armory of knowledge from a time beyond the reach of estimate. To be sure, spinners of tales might leap over the fence of reality into realms of phantasy where dragons would grow from innocuous seeds and beanstalks would defy all restraints and reach to the sky. But the farmers of antiquity planting their grains in the valley of the Nile, the Euphrates and the Yangtze confidently expected to harvest not dragons or monstrous beanstalks but multiple replicas of their precious, planted grain.

Even more subtle expressions of biological inheritance were understood by primitive peoples. Arabian racehorse breeders kept elaborate pedigree records of their racehorses for centuries. But while it was recognized that certain traits which can vary among individuals are passed down from parent to offspring, the patterns which heredity follows were unknown until very recently.

It is odd how knowledge of ourselves consistently lags behind knowledge of our outer world. Johannes Kepler perceived the orbital motion of the planets over 350 years ago. We have confirmed his hypothesis only in the past few years when we achieved man-made, unfettered orbital motion. Yet this remarkable genius who, with rare perception, could bring order out of the apparent chaos of celestial bodies could not escape the prison of ignorance in contemporary biological lore: He believed that fish could arise by spontaneous generation from the salt water of the seas just as comets arise in the skies. And in the latter half of the nineteenth century, when we were the masters of the seas with palatial steamships, we could not even guess at the probability of the birth of a blue-eyed baby to brown-eyed parents.

For example, Darwin, whose hypothesis on evolution was based on the assumption of the existence of heritable variations among individuals of the same species, knew nothing about the patterns that heredity follows. In 1872 he wrote:

The laws governing inheritance are for the most part unknown. No one can say why the same peculiarity in different individuals of the same species, or in different species, is sometimes inherited and sometimes not so; why the child often reverts in certain characteristics to its grandfather or grandmother or more remote ancestor.

Ironically, the answers to the questions Darwin was seeking had been known for six years. But the work of the brilliant amateur biologist who observed the pattern—indeed the laws—which heredity follows lay unread and unappreciated in a dusty volume of the *Proceedings of the Natural History Society of Brunn*.

The obscure biologist was Gregor Johann Mendel, who was born in 1822 in Silesia to peasant parents. After finishing his secondary education he endured economic and physical hardships and decided to enter a profession in which "he would be freed of the bitter necessities of life." Thus, at the age of twenty-one, he became a monk.

After his ordination he was assigned to duty as a temporary

teacher and remained in this category because he kept failing the examinations for a full-fledged position. It has been suggested that Mendel was stimulated to start his experiments by a dispute with his examiner in botany on the last of his unsuccessful examinations. Whatever were his motivations, he pursued his studies on plant hybridization in the tiny garden of the monastery of St. Thomas at Brunn, Austria, with such devotion and brilliance that he was lifted from his self-imposed anonymity and is counted among those men of rare genius who are the first to discover a law of Nature.

He studied the effects of the crossbreeding of two plants endowed with contrasting traits. He started by planting different peas, tall and short, in the monastery garden. If these two varieties of peas were allowed to self-fertilize, with no possibility of cross-fertilization, the plants yielded seeds which bred true to type: they grew into plants, tall or short, like their parents. But if a tall and a short pea were crossed-fertilized or "hybridized," the seeds from such a union gave rise only to tall peas. Mendel did not stop there. He patiently crossed these tall peas of mixed ancestry and collected their seeds. Out of these seeds grew both tall and short peas. He carefully recorded his seeds and crops and found from over a thousand different plantings that the tall and short "grandchildren" always appeared in a definite ratio: 75 percent tall and 25 percent short. He repeated these studies with red- and white-flowered peas. When he cross-fertilized these flowers he found that all the seeds gave rise to red flowers. But this new generation of red flowers, when cross-fertilized, produced seeds from which grew both red and white peas. He found 6022 red and 2001 white flowers, again a ratio of three to one (75.1 to 24.9 percent, to be more exact).

Mendel concluded that there are factors in peas which determine their color and height. The factor for whiteness or shortness remains dormant in the first generation after the crossbreeding of opposing traits, but asserts itself in the second. Moreover, the dormant factor reappears in the second generation in a ratio of one to three of the dominant factor.

So here for the first time the pattern of inheritance was revealed. Characteristics of the two parents are not transmitted to the offspring haphazardly but by some directing mechanism which achieves sufficient accuracy to entitle the pattern to be called a law.

One looks upon Mendel with admiration and envy. His experiments were designed with beautiful simplicity and carried out with impeccable technique. His selection of the sweet pea was no accident. He gave his reasons for the choice: Peas were available in many pure breeding varieties; the flowers were protected anatomically from the danger of intrusion of foreign pollen; and, finally, plant fanciers knew the hybrid varieties to be perfectly fertile.

With the intuition of genius he chose to work with sufficiently large numbers so that statistical accuracy could emerge from them. His conclusions he drew with bold incisiveness.

Our envy is directed toward his blessed status as an amateur. The amateur has vanished from the field of science. Mendel was perhaps the last of the great ones. Today a scientist must be a small-time entrepreneur whose concern with budgets, grant applications and progress reports at least equals his preoccupation with science. And above all, today's scientist must publish rapidly to survive.

Mendel was aware of the importance of his discovery. He tried to get fellow scientists interested in it and sent a copy of his findings to an outstanding Swiss botanist, Karl von Nägeli. But the latter had his own ideas on the mechanism of heredity and brushed aside the presumptuous claims of an obscure amateur.

So the meticulous report on the results of Mendel's eight years of work was buried in the pages of the provincial journal where they were printed in 1866. Two years later Mendel was elected the abbot of the monastery and, as has happened to a good many scientists since then, he abdicated science and became an administrator. He died in 1884 completely neglected by the scientific world which was to discover him only sixteen years later.

This lack of recognition should not impel us to think of Mendel with pity. He was one of those fortunate few among men who was permitted to do what he wanted to do. We get from one of his letters a glimpse of him at work.

As must be expected the experiments proceed slowly. At first beginning some patience is required, but later, when several experiments are progressing concurrently, matters are improved. Every day, from spring to fall, one's interest is refreshed daily, and the care which must be given one's wards is thus amply repaid.

Mendel distilled the essence of the life of the real scientist into one sentence: "Every day from spring to fall one's interest is refreshed daily, ... "

This then is the ultimate reward of the scientist; not power, not a professorship, not bigger grants, but the complete immersion in work which buoys him with interest and gives, if not an assurance, then at least an illusion of being worthy, real, and lasting.

The discovery of Mendel's work was made simultaneously by three different investigators who through studies of their own arrived at the same conclusions as did the patient monk.

The three, Hugo de Vries, who was a Dutch botanist, Carl Correns, a German botanist, and Erich von Tshermak, a Viennese plant breeder, apparently learned of Mendel's work from a reference in a comprehensive bibliography on plant hybridization compiled in 1881 by some meticulous German scholar. Each of them graciously acknowledged Mendel's priority on the discovery of what they designated as "Mendel's laws."

What were these laws? In the first place Mendel determined that a single pollen achieves fertilization. (This is, of course, also true of fertilization among animals where only one sperm can penetrate an egg.) Mendel set the pattern for studying the paths of heredity: One must choose a single pair of easily recognizable, contrasting characteristics, e.g., tall, short. One of these may turn out to be a *dominant* and the other a *recessive* trait.

Recessive traits vanish from the appearance of the second generation, only to reappear in the third generation in a ratio of 1 recessive to 3 dominant.

Finally, Mendel assumed the existence of a "formative element" in each pollen and each egg which is capable of determining a single character, e.g., shortness or whiteness, in the offspring.

Mendel's experiments have stood the test of countless repetitions with every species of living organisms that reproduce by the fusion of two sex cells. Every creature from man to mouse shows recessive and dominant traits and the expression of these traits usually follows Mendel's law.

During the hundred years following the publication of Mendel's findings we have slowly pieced together the molecular mechanisms which unerringly achieve the transmission of heritable traits to the offspring.

The pioneer who unwittingly made the first contribution toward the unraveling of the molecular mechanism of inheritance was the Swiss scientist Friedrich Miescher. His work was contemporaneous with Mendel's and without realizing it he isolated for the first time the component of the cell which turned out to be the "formative element"—the hypothetical substance to which Mendel intuitively ascribed the capacity to express a heritable trait.

The second half of the nineteenth century was an era of golden harvest in the biological sciences. The first half was an era of groping exploration. In 1828 Wöhler synthesized in the laboratory urea, which theretofore had been made only by living organisms. He thus broke down the walls between the chemistry of living and nonliving worlds. With this finding he assaulted the principle of vitalism, which held that living organisms were endowed with mysterious vital forces that were beyond the reach of rational inquiry. Only ten years after Wöhler's discovery Schleiden formulated the cell theory, stating boldly that the embryo of a plant is a single cell and the growth of a plant results from the continuous division of such cells. Thus the cell was pinpointed as the unit of life and the defeatist taboo against its chemical exploration was successfully defied by Wöhler. A botanist at the University of Tübingen investigated the contents of the cell and called the gelatinous substance encased by the cell wall protoplasm. Karl von Nägeli, who was blind to Mendel's profound discovery, redeemed himself by the discovery that protoplasm is a nitrogenous substance. Protoplasm came to be looked upon as the physical matrix of life and scientists began to explore it with every tool at their disposal.

Laboratories for scientists began to be built at universities and the spread of the rapidly accumulating knowledge about our biological world was facilitated by the founding of specialized journals. Intellectual isolation was thus slowly broken down, and with the greater ease of travel students flocked to leaders in the field of their interest wherever they were.

Friedrich Miescher of Basel, Switzerland, was such a traveling student. He was interested in the makeup of the nucleus within the protoplasm. He thought looking at a cell's nucleus through a microscope, after a variety of stains were applied to the tissues, was not enough, so he traveled to Tübingen, Germany, to study with Professor Felix Hoppe-Seyler, who was an expert in what was then called medicinal chemistry.

It was known from microscopic examinations that the nucleus made up an inordinately large proportion of a pus cell. So Miescher, with Swiss patience, started to collect the oozing surgical bandages which were peeled off purulent wounds in the hospital of Tübingen. He soaked off the organic material, digested the product with the digestive enzyme of the stomach, pepsin, and subjected the mixture he obtained to a variety of chemical manipulations. That he did not have an easy time of it was obvious from his contemporary comment: "I feel as if I am mired in a swamp."

Finally his exertions yielded a material which, unlike other proteins then known, was not soluble in water or dilute acid but did dissolve in dilute alkali. Its difference from other proteins was marked even more when it was found to contain phosphorus. This element had until then been only found in one other constituent of the cell: Hoppe-Seyler himself found it in a component of fats.

Miescher named the novel substance he found in nuclear material "nuclein" and wrote up his efforts in a scientific paper and sent it in 1869 to Hoppe-Seyler to publish. That gentleman edited a journal, which he had named in a burst of modesty *Hoppe-Seyler's Journal of Medical Chemistry*. Hoppe-Seyler was a suspicious editor and would not publish Miescher's paper until he himself had repeated and confirmed the surprising discovery. So two years later Miescher's original account, plus Hoppe-Seyler's confirmation, plus two new papers on the same subject by two of Hoppe-Seyler's students appeared simultaneously.

These uninvited fellow travelers of Miescher's paper showed that nuclein is not restricted to pus cells but is present in the red cells of animals, in yeast cells, and in casein from milk.

This overcautious editorial policy with a bit of piracy on the part of Hoppe-Seyler's two students would not be tolerated today, but it did set the pattern of procedure for scientific journals.[1]

[1] Hoppe Seyler's cautious editorship did not always have such a happy outcome.

In the 1880s the Scotch scientist MacMunn made an astute observation on the presence of hematin (a pigment similar to that of the red pigment in hemoglobin) in tissues other than blood. When these studies were submitted for publication Hoppe-Seyler peremptorily rejected them with the suggestion that MacMunn was dealing with contaminating pigments from blood. It was

Today before a scientific communication is accepted for publication in most journals it must first be approved by two anonymous experts in the field who can attest from their experience that the findings reported by the author are new, not too trivial, and plausible.

Miescher returned to his native Basel, where he hit upon a more congenial source material for his continuing studies of nuclein.

Basel is on the river Rhine and salmon battle their way up this river to their fresh-water spawning grounds. The sperm of salmon is hardly more than a nucleus with an apparatus of locomotion, the tail, attached to it. Salmon sperm became the main source for Miescher's studies—a welcome change from the stinking surgical bandages of Tübingen.

While studying the sperm of salmon Miescher made another penetrating observation. Salmon do not eat during their fresh-water pilgrimage to their spawning grounds. (Their stomachs were empty of residual products of digestion.) During the voyage the musculature of the salmon diminishes while, concomitantly, their sperm content burgeons. Miescher suggested that in these fish muscle must be converted to sperm. In view of the primitive state of knowledge about body chemistry at that time this was a prophetic observation on the interconvertability of body constituents. Not until half a century later could we prove conclusively by means of isotopic labels the validity of Miescher's penetrating conclusion. He intuitively sensed the importance of nucleins. He wrote:

A knowledge of the relationships between nuclear materials, proteins and their immediate products of metabolism will gradually help to raise the curtain which at present so completely veils the inner process of cellular growth.

---

not until 1925 that MacMunn was vindicated when the Polish scientist Keilin, working in England, showed that not only were these pigments real, but were important components of the cell's respiratory mechanism. In this case Hoppe-Seyler, with his highhanded editorship, hindered the growth of biochemistry by decades.

So he kept studying his nucleins with persistence and high skill. His analyses of their elementary components would be acceptable today.

He suggested that a study of the structure of nucleins from different sources might lead to a discovery of characteristic differences among them. He submitted this suggestion for publication to Hoppe-Seyler, who left us one more example of his stern editormanship by refusing to publish it, and Miescher's suggestion came to light only after his death at the early age of fifty-one.

After Miescher one of his successors in the field coined a new term which competed with "nuclein" so successfully that the latter retains only a historical interest whereas "nucleic acid" today is almost a household word. For by now we know that every living organism from the mighty whale to the tiniest speck of virus particle is built from a blueprint of nucleic acid: The secret of the mechanisms of heredity whose pattern was first glimpsed by Mendel is locked in the wondrous structure of the substance first isolated by Miescher.

Nearly a century of scientific effort by those who followed Mendel and studied whole organisms, and those who followed Miescher and studied extracts of organisms, was needed to achieve this fusion of the work of two pioneers. It was a confused century in other ways, producing as it did the most monstrous wars, fought with weapons of grisly horror. But the biological scientist can look back a hundred years to Mendel and view with pride and tingling excitement the steady path—sometimes winding to be sure—that we have followed. A thousand years from now historians—if there will be such—may mark this century not as the one that yielded the most monstrous slaughter of man or the one in which we unleashed the energy of the atom, but the century in which we deciphered the Code of Life.

2 THE UNIT OF LIFE, THE CELL

AND THE SITE OF HEREDITARY CHANGE

AFTER the pioneers came the developers. The earliest successes were scored by those who studied intact, living organisms. This was to be expected; for did not Mendel get much farther than Miescher? A certain amount of initial success is inherent in classical biological experiments. Little is needed in the way of elaborate equipment or refined techniques. The subject of the study, the living organism, does the work. The scientist needs curiosity, patience, the clarity of mind to ask questions that are simple but penetrating, and, finally, he has to have the humility to accept the answers as given by Nature. Both Mendel and Nägeli were working on hybrid formation in plants, but when Mendel sent the latter one of most penetrating set of observations ever obtained from a single experiment Nägeli rejected them because the sweet peas did not conform to the dogma which Nägeli had ordained for them. He wrote to Mendel: "You should regard the numerical expressions as being only empirical because they cannot be proved rational."

But others, of humbler nature and more receptive mind, saw

a tiny part of the mechanism of life suddenly illuminated by the flash of genius of the gentle monk.

The man who contributed more than anyone else to the development of the frontier to which Mendel's work guided us was an American, Thomas Hunt Morgan. He was a Southerner —his uncle was the Morgan of Morgan's Raiders—who took his doctorate in biology at Johns Hopkins. Involved in a variety of biological problems and actually critical of Mendel's conclusions, he was designing experiments to challenge them. He decided to try to induce mutations and chose as his experimental organism the fruit fly, which had already been introduced as a tool for biological studies. The fruit fly (*Drosophila melanogaster*) offers many advantages as an experimental animal, not the least of which is the ease of its collection. One merely leaves a piece of overripe banana on a laboratory bench and the fruit flies will swarm over the fermenting mound. A rare, white-eyed fruit fly alighted on a laboratory bench in the Zoology Department at Columbia, where Morgan was a professor. It was noticed by Morgan's wife, who was also a biologist, and after a hectic hunt by a posse of students the hapless visitor was captured alive. Morgan induced this white-eyed female—for such it turned out to be—to mate with a red-eyed male and studied the eye color of the offspring of the union. The eye coloration of succeeding generations did not follow the simple pattern expected from Mendel's laws. This unexpected anomaly became a test of Morgan's character and intellect. A lesser man might have been satisfied with this challenge to Mendel's original conclusions and reaped the transient notoriety of the wrecker of one of the recently erected pillars of science. But Morgan decided to continue to pursue the subject and attempt to reconcile the anomaly with Mendel's laws. To do this he summoned all his knowledge of recent advances in biology, especially of cytology, a subdiscipline which was just about then groping toward maturity.

The science of cytology encompasses the study of cells from every conceivable source under every conceivable condition.

Cytologists are the unsung heroes of biological science. Their work is not spectacular enough to invite dramatization by those popularizers of science who feel that the substance of science is a pill to be fed to the reader with a sugar-coating of real or manufactured drama. Cytologists do not design dynamic experiments. They merely sit at a microscope and observe with infinite patience cells exposed to a large variety of chemical dyes. It is a visual exploration of a miniscule universe, for we must recall that a cell is so small it would take 5000 human red cells to cover the dot over a letter "i" on this page. Yet this tiny, jelled, watery world contains a wondrous variety of structures. Among these are the chromosomes, which are rod-shaped bodies in the nucleus which take on basic dyes more readily than does the rest of the cell. (This propensity for being dyed is, of course, the origin of the name: pigmented bodies.)

The American cytologist C. E. McClung noted that the male grasshopper has an odd number of chromosomes whereas the female has an even number, and he concluded that the extra chromosome determines the maleness of its owner. This was an extraordinarily bold conjecture, for McClung was pinpointing the structure in the cell which accounts for the difference between a male and a female and thus was assigning a role to the chromosome as the agent which determines such a heritable trait as sex.

Morgan was aware of McClung's hypothesis and with a flash of intuition associated it with his anomalous findings on the inheritance of eye color. If the inheritance of the eye color in the fruit fly were in some way linked to the sex chromosome, then the absence of the latter from one half of the population would vitiate Mendel's simple statistics which had been obtained from studies of a plant whose chromosomes are evenly distributed to the offspring. Morgan followed up his initial observation with driving energy and penetrating intellect. His choice of the fruit fly was very fortunate, for this animal breeds rapidly (ten days to two weeks for a generation), it has many offspring, which makes studies statistically valid, and has a small number of chromosomes, only four pairs. Morgan's studies of the fruit fly laid the solid foundation for our knowledge of genetics.

As Morgan and his students were unraveling the hereditary maze of the fruit fly they found that other traits besides the eye color and sex are linked together and four such groups of linked traits were identified.

Since there are more traits than chromosomes, and since several traits are linked together, the conclusion was drawn that one chromosome houses the controlling mechanisms for the linked traits. These tiny, invisible fragments of the chromosome, each of which controls a single trait, are now called genes. It is believed that a separate gene exists for every single trait.

How do the genes dominate the cell? And, since an organism is an aggregate of cells, how do the genes determine the shape, structure, and very being of each organism? T. H. Morgan has said: "Cytology furnishes the mechanism that the experimental evidence [from genetic studies] demands." Let us therefore turn to the cytologist's view of the cell.

Our image of the cell has varied with the instruments available for enlarging our field of vision. Indeed, the microscope is but a by-product of early attempts to come to the aid of the human eye, that organ so wondrous in its abilities, so devastating in its shortcomings. Knowledge of the optical properties of curved transparent surfaces reaches back over two thousand years, but the mustering of that knowledge to aid our vision had to wait until the end of the thirteenth century when the Italian Salvino degli Armati invented spectacles. As a consequence, lens grinding became a widely practiced craft, and the microscope is but a serendipitous by-product of that profession. In 1590 two Dutch artisans of the optical industry, Zachary and Francis Janssen, combined two convex lenses in a tube and improvised the first compound microscope. With their primitive instrument they opened a window on a universe hidden from the limited range of the unaided human eye. Seventy-five years later Robert Hooke first observed the cellular structure of the bark of a tree. (The word for the study of cells, cytology, stemmed from the Greek *kytos*, meaning hollow, empty space. The derivation of the word is a symptom of the fallacy in the observation of Hooke and many who followed him. They focused so hard on the cell walls, that they neglected their content.) Hooke meticulously sliced pieces of cork and examined the thin wafers with the aid of a primitive microscope. He presented what he saw before the Royal Society of London in patiently drawn diagrams. As he described them: "The *cells* were not very deep, but consisted of a great many little boxes, separated out of one continued long pore, by certain diaphragms." (We can see Hooke's drawing in Figure 2.1.)

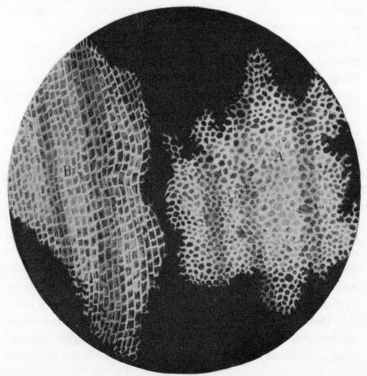

Fig. 2.1. Hooke's drawings of cork cells

As often happens in the growth of knowledge, many workers added bits of information on the cellular composition of living tissue and from the very diversity of observations confidence in its reality grew.

Those who like their history in concentrated form credit Schleiden (1838) and Schwann (1839) as the originators of the cell theory, but actually they merely harvested the bloom of evidence which grew out of the patient work of scores before them. For example, in 1809, some thirty years before the pronouncements of Schleiden and Schwann, Jean Baptiste Lamarck stated: "No body can have life if its constitutent parts are not cellular tissue or are not formed by cellular tissue." Schwann stated it more categorically: "The cells are organisms arranged in accordance with definite laws."

During the early part of the nineteenth century, as the contents of the cell began to engage the attention of investigators, the term protoplasm was coined to describe the mucilaginous material which filled the cell. Unaided by the staining techniques which later revealed the complexity of the cell, the early workers saw, in the words of the French biologist Felix Dujardin in 1835: "A perfectly homogeneous, plastic contractile, diaphanous gelatinous substance, insoluble in water and without traces of organization." This statement, as we shall see, turned out to be *the* misstatement in the history of cytology. The cell is an exquisitely structured, faultlessly organized, self-contained world. The topography of this Lilliputian universe emerged slowly as a result of the cumulative effort of generations of cytologists who kept improving with patient inventiveness their tools and techniques. Machines were designed for slicing tissues to gossamer thinness; dyes were tested by the hundreds in hopes that they would reveal, by selective staining, structures hitherto unobserved. The scrutiny of the prepared slides required imagination and objectivity in equal measure. Is the tiny speck appearing under the microscope a real structure or is it an artifact of the preparation? The latter unfortunately abounds whenever the highly organized system of tissues and cells is disrupted. The literature of classical biology has copious descriptions of such structural artifacts. Nor was preoccupation with artifacts restricted to the classical biologists. Currently, as biology has moved down to the molecular level, the literature abounds in molecular artifacts. We discover chemical activities in isolated systems which have no counterpart in the intact organism. Molecules released from the indigenous structures of the cell escape the many built-in regulatory controls and exhibit functions in the test tube that have no relation to the living process.

While artifacts are numerous, actual chicaneries—the novels of C. P. Snow to the contrary notwithstanding—are extraordinarily rare in science.[1] This may be due to some selective process among those who choose science as a career, or, more likely, it is the result of the restraining influence of a tradition of critical evaluation and repetition of experimental claims. Every scientist worthy of the name is well trained in the arts of a devil's advocate. Science is the only creative endeavor that needs no professional critics: everyone of us assumes that role. And that is as it should be. The imaginative scientist spends much of his wakeful hours in a world of phantasy. In a fertile mind a microscopic smudge or a click of a Geiger counter is amplified into a

[1] The extraordinary rarity of conscious fabrication of data by scientists can be gauged by the fact that in the past twenty-five years the writer has heard of only three such incidents among American biochemists. The reaction to these instances of perfidy—perpetrated by fledglings in the field—was a profound sense of revulsion, as if an odious crime against someone loved and respected had been committed. The degree of resentment indicated that the crime penetrated deeper than a mere breach of the rules of the game. By fabricating nonexistent phenomena for their own advantage the miscreants were attempting to counterfeit a small part of Nature itself. It was thus a profanation of the only thing many of us hold sacred, the splendor of Nature's ways.

beacon which illuminates a tiny segment of the intricate maze that is a living cell. Is this vision real or is it an artifact elevated to reality by an overambitious phantasy? A conscientious scientist invites challenge of his work at the numerous meetings that are organized just for such purpose. Only from the survival of the challenge of our fellow craftsmen do we gain assurance. Only work and conclusions which are constantly exposed to the open, international courts of scientific inquiry gain credence. Isolation and secretiveness convert laboratories into artifactories.[1]

The fruitful imagination of generations of cytologists and their skill with slicing and staining for the visual microscope yielded by the early quarter of this century the image of the cell which is shown in Figure 2.2.

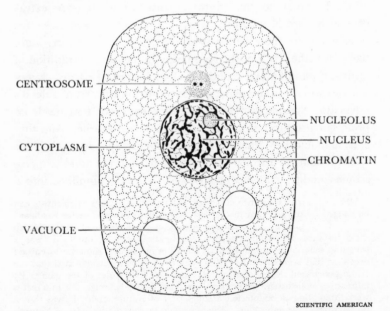

SCIENTIFIC AMERICAN

Fig. 2.2. A typical cell as seen under a light microscope in the early decades of this century

But even greater magnifications, and therefore clarity, were to emerge when the electron microscope was focused on the cell. This instrument threw off the shackles imposed by the limitation inherent in the optics of curved surfaces. The electron microscope channels onto its target streams of electrons which are constricted to narrow piercing beams by strong magnetic fields. Instruments in current use achieve magnifications about five hundred times greater than the visual microscope.

In fairness to the generations of discerning cytologists who had been peering through visual microscopes, very little in the fundamental structure of the cell which had eluded them was revealed by the electron microscope. When one walks into a strange mountain range at night one can discern by the light of a pale moon much of the encircling topography. But as the rising sun splashes its rays, the same area presents a new, sharp aspect. Dim, smooth shapes leap into clarity as tortured surfaces crisscrossed by gulleys and crevasses. The dark, homogeneous, wooded slopes of the night reveal a surface of lush vegetation varied in size, shape, and foliage. So a new image of the cell leapt at us when we focused beams of piercing electrons instead of the rays of visible light on it. Such a view of the cell taken with an electron microscope is seen in Figure 2.3.

Let us look at the structures of such a cell in detail. The cells of plants and of bacteria are enclosed by semirigid cell walls, but animal cells are not endowed with this armor.[2]

All cells, whether of plant, animal, or bacterial origin, are completely enclosed by a membrane which guards the physical integrity of the cell, maintaining it as an inviolate semiautonomous unit. Not only does the membrane guard the cell by

[2] This anatomical difference is fortunate, for it provides the basis for the most effective protection of ourselves against bacterial parasites. The point of attack by some of the most effective antibiotics against bacteria is the cell wall. Penicillin, for example, prevents the formation of bacterial cell walls. Since the animal hosts do not possess this structure, penicillin is selectively destructive of bacterial cells only.

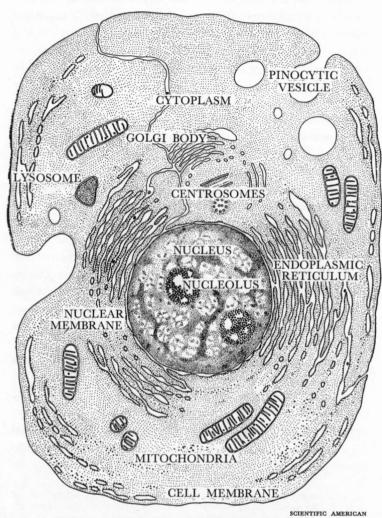

Fig. 2.3. A typical cell revealed by an electron microscope

physical containment but it also mounts an unceasing guard over the ports of entry on the frontiers of the cell. The membrane recognizes with its uncanny molecular memory the hundreds of compounds swimming around it and permits or denies passage according to the cell's requirements. Unknown compounds which fail some subtle test for recognition on the molecular ramparts of the membrane are usually excluded and the cell is thus guarded against their possibly harmful presence.

The cell's membrane girds a kingdom of a topography so varied it would astonish the good Doctor Dujardin. Some of these structures are obligatory. These are the organoids, which must be present in every cell of a given tissue. There are other structures whose distribution is more random in otherwise homogeneous cells. These are the inclusions which may be a tiny blob of fat or a grain of starch. Organoids must be reproduced during cell division; inclusions need not be. The function of some of the organoids in the cell is well known; others still manage to hide their activities.

Our knowledge of the function of the organoids depends on our ability to remove them in more or less intact form from the cell. In order to achieve this the integrity of the cell must be destroyed. This may be achieved by mechanical grinding with some abrasive powder; or it may be done by exposure of the cells to sonic vibration; or in the case of bacteria, their cell walls can be dissolved by an enzyme, and without its protective enclosure the cell pops and disintegrates like an overblown balloon. The next step involves a series of centrifugations at higher and higher speeds and, consequently, at gravitational forces of increasing intensity. Exposing the cells' contents to 100,000 times the force of gravity is a routine operation in most biochemical laboratories today. The various components of the cell sediment out sequentially as increasing centrifugal fields are imposed upon them. The many different fractions are harvested individually; their appearance is examined under an electron microscope; visual correlations are then attempted with the organoids as they appear in the intact cell. With luck we can obtain preparations of some of the organoids that retain some of their original functions. For example, there is a widely distributed structure which is just barely visible under an ordinary microscope. At the present time it bears the name mitochondrion. (The name is a remnant of the confusion about this organoid: it means a filament [*mitos*] and a granule [*chondros*].) However, since its original discovery in 1894 it has borne no less than fifty different names given by as many cytologists, each one of whom claimed it as a novel

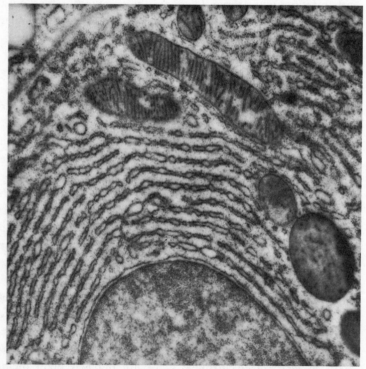

JOURNAL OF BIOPHYSICAL AND BIOCHEMICAL CYTOLOGY

Fig. 2.4. The slipper-shaped structures are mitochondria

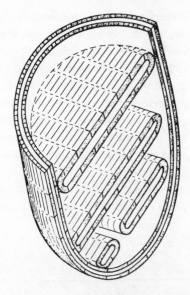

Fig. 2.5. A three-dimensional construction of a mitochondrion

discovery of his own. Under the electron microscope the structure which sometimes appeared as a granule and sometimes as a filament emerged as a highly organized unit with a characteristic anatomy of its own. The reader can see an electron photomicrograph of a mitochondrion in Figure 2.4. We can get an approximation of their real size if we attempt to visualize a thousand such mitochondria forming a ladder across a dot on a letter i. Yet this tiny component of the cell has an elaborate structure, which is emphasized in the idealized drawing of Figure 2.5.

The function of the mitochondrion emerged in the early 1950s as a result of two entirely different lines of investigations. On the one hand there were investigators interested in the electron microscopic anatomy of the cell who were concentrating the mitochondria by differential centrifugations. About the same time, biochemists who were interested in the mechanisms with which the cell generates its energy from glucose were concentrating particles from disintegrated cells that could carry on oxidation even outside the cell. When the two groups, the anatomists and biochemists, compared their preparations they found them to be the same: The mitochondrion turned out to be the furnace of the cell. It is a furnace with at least forty different working enzymes embedded in it, probably in fixed positions. The enzymes, like so many workers on an assembly line, oxidize glucose and store its precious energy into a form readily usable by the cell for a multitude of tasks that require energy.

Another organoid whose function is known is the chloroplast, which is found in plant cells and in some bacteria (see Figure 2.6).

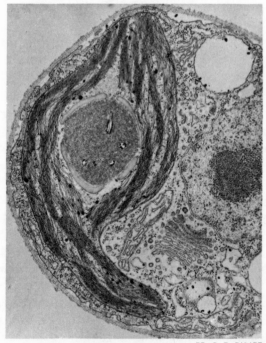

DR. G. E. PALADE

Fig. 2.6. A chloroplast

The function of the chloroplasts was relatively easily deduced. The telltale green color of the preparations indicated the presence of chlorophyll, the catalyst of photosynthesis. Thus, chloroplasts are the structures on which all forms of life eventually depend. They alone have the life-generating ability to concentrate the sun's prodigious but diffuse energy into a form which can be the fuel of life. With extraordinary efficiency they harness the electromagnetic energy of light and use it to pack random, disorganized molecules of carbon dioxide into the highly organized, energy-rich structure of the glucose molecule. In turn, the enzymes in the cells of every living organism can reverse the process and tap the energy within the glucose molecule for their own needs as they dismantle it down to carbon dioxide. Living organisms can perform various transformations of energy. The ear converts sound, and the eye converts light to electrical energy. The skin can translate mechanical pressure to an electrical signal, and the nerve can transform chemical to electrical energy. Some organisms such as the firefly can use their stored chemical energy to generate light. Our muscles consume chemical energy for motion; our vocal chords do the same but produce an exquisitely controlled motion which gives rise to sound. The source of energy for all of these transformations is the specialized molecular battery adenosine triphosphate (ATP), which is produced in the mitochondrion from the chemical energy of the glucose molecule. In turn, it was the chloroplast which, with virtuoso skill, trapped the energy of light and transformed it into the energy that holds the glucose molecule together. The writers of the Old Testament, with perceptive intuition, placed the turning on of light very early on the agenda of the task of creation. Light is the bountiful seminal source of all life and the chloroplast is the womb which reshapes light into the stuff of life.

Another organoid whose function is well understood at the present time is the ribosome, which is the site where protein molecules are assembled from the appropriate amino acids. We shall devote a whole chapter to this process, and therefore the detailed description of the ribosome can be postponed until then.

Although we have been studying the cell intensively for 130 years, we can still discover new structures in that wondrously complex speck of life. The lysosome is such a recently identified unit; it approximates the mitochondrion in size but lacks its highly organized structure. The lysosomes seem to be bags of digestive enzymes that can dismantle large protein and nucleic acid molecules. The rubble of small fragments so produced can pass through the membranes of the mitochondrion and be consumed in its furnace. It is obvious why such potent agents within the cell must be contained, otherwise the contents of the lysosomes would destroy the cell by literally boring from within. What the role of these enzymes is within the cell or how the structures that are doomed for dismemberment pass into the lysosomes is obscure at present. This may be the method of elimination of defunct or excess structural or functional components of the cell. Studies with isotopic tracers have shown that the molecular components of all living things are in a constant state of flux; tissues are being built up and broken down simultaneously. The molecules which compose our bodies today will be gone in a few days and replaced by new ones from our foods. Even highly organized structures such as the mitochondria have but a transient life within the cell. The life span of a mitochondrion of the liver has been estimated to be from ten to twenty days. The reason for this profligate discarding of cellular components is not known. It is apparently easier to assemble a brand-new mitochondrion than to repair one whose structure has developed some molecular fissure or whose functional efficiency has faltered. An uncannily perceptive method of recognition must single out the defective structure for dismemberment by the enzymes of the lysosomes, and the resulting structural debris is fed in as fuel to the furnace of intact mitochondria.

The Golgi bodies, which were discovered in 1898 by the Italian whose name they bear, are still another example of well-defined organoids (Figure 2.7). However, despite their venerable history and ubiquitous distribution in animal cells, essentially nothing is known of their function. One hypothesis is that the Golgi bodies are the sites where some of the membranes for the cell are assembled. However, this must remain as an unverified conjecture until we can isolate these structures in concentrated form and induce them to perform their tasks—in the jargon of science—*in vitro*.

What other organoids may be in the cytoplasm but too small for recognition with our present methods of observation, we do

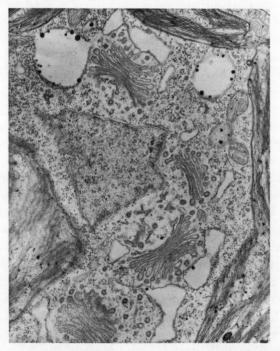

DR. G. E. PALADE

Fig. 2.7. The tubules bunched together like sheaves of wheat are Golgi bodies

not know. Faultless structure is the hallmark of the cell and we can therefore predict that in the range between single molecules and relatively giant structures such as the visible organoids there must be intermediate organizations designed for some special tasks. The detection of these "micro-organoids" will be the task of future cytologists. The writer's confidence that such structures must exist stems from our knowledge that even the single protein molecules of the cell have elaborate three-dimensional structures. One such molecule whose structure and function are both well known is hemoglobin. Since it can be isolated chemically with relative ease, its function was deduced in the early morning in the history of biochemistry. Our knowledge of its three-dimensional structure has evolved from the painstaking work of biophysicists. Since the molecule is beyond the reach of even the electron microscope, an entirely different approach to the decoding of molecular shapes had to be devised. At the present moment a technique called X-ray crystallography gives the greatest promise of revealing the structure of the cell's macromolecular components and ultimately of the structure of giant molecular constellations within the cell. Let us therefore digress into a brief history of this technique.

In 1895 Röntgen shot a stream of electricity onto a metal target in a vacuum tube and discovered that the impact gave rise to a new kind of light—a light that penetrated some opaque barriers which shut out visible light. The English physicists J. J. Thompson and W. H. Bragg took up the study of these puissant rays and the former came to the conclusion that he was dealing with waves of extremely short wavelength—on the order of 10^{-8} or 10^{-9} cm. Bragg, on the other hand, showed that X rays behaved like discontinuous bundles of surging particles. Of course, the two of them foreshadowed the modern quantum theory, which reconciles the peaceful coexistence of particulate and wavelike attributes for electromagnetic radiations. The

German physicist Max von Laue suggested that, since the wavelength of X rays was claimed to be shorter than the distance between the parallel rows of molecules in crystals, different crystals might be used to probe the nature of X rays. He visualized some of the rays getting deflected as they hit the single molecules in a crystal. And just as evenly placed stone pillars in water cause perturbations in the waves passing through them, so von Laue had anticipated similar diffraction patterns to emerge as X rays passed through the evenly spaced, miniscule molecular abutments within a crystal. Painstaking experiments soon confirmed the validity of von Laue's hypothesis. When X rays are passed through different crystals, beautiful, symmetrical patterns of diffraction spots appear on photographic plates placed beyond the crystals in the path of the rays. Thus the crystals probed the nature of X rays. The reverse and greater potentiality, the probing of the nature of crystals by X rays, did not seem to occur to von Laue. But fortunately, by 1912 communication among scientists was easy.[3]

W. L. Bragg, a young Englishman of twenty-two, was interested in crystal structure. His father, W. H. Bragg, a professor at the University of Leeds, was an expert on the characteristics of X-ray tubes and the rays they produced. The two of them joined forces and laid the foundations for X-ray crystallography,

[3] To be sure, exchange of ideas was to be shut down two years later by World War I, which was to waste the minds and bodies of those who would have shaped science. One of the most brilliant young Englishmen to study the properties of X rays, H. G. J. Moseley, was killed in 1915 in the fatuous Gallipoli campaign. Before his death at the age of twenty-eight, he had established a solid reputation as a scientist of first rank. It was Moseley who determined that the charge on the nucleus of the atom goes up in regular increments with increasing weight of the atom. He did this by examining the nature of the X rays produced when each element is made the target of a beam of electrons. He demonstrated a physical basis for atomic numbers which up to then were a mere incidental numerical attribute derived from increasing atomic weights of the elements. Contemplation of what else Moseley might have achieved is both awesome and jejune.

the science of determining the structure of matter from the kind of target it offers to X rays. For their work father and son were honored in 1915 with a joint Nobel Prize.

After the war the two Braggs became geographically separated and they staked out different areas for exploration. The son, W. L., was at the University of Manchester and he and his students attacked the structure of inorganic crystals of increasing complexity. The father, W. H., moved to the University College in London and focused on the structure of organic molecules. He trained several young people who in turn staked out wider and wider areas for exploration. Among these, William Astbury had the courage to undertake the study of such an amorphous, complex natural product as wool under conditions of varying temperature, moisture, and mechanical stress. (Astbury had an impish sense of humor. During a lecture before New York's learned Harvey Society he flashed on the screen what he called the most beautiful example of stretched wool. It was a thin sweater, snuggly hugging a lady of sufficiently ample development to qualify her as a star in Hollywood or Rome.)

Astbury's work pioneered the study of the structure of such complex components of the cell as a protein molecule or a nucleic acid. After decades of work we recently had unveiled to us the three-dimensional shape of two such molecules. One of them, that of hemoglobin, can be seen in Figure 2.8. This is but one of the thousands of different species of molecules within the cytoplasm of a cell whose function it is to supply its every need and permit it to perform its ultimate function: to divide and proliferate. For there has been a relentless pressure from time immemorial, within every cell, to surge toward growth and duplication and thus ensure the continuity of life.

The apparatus for the achievement of that continuity is invested in most cells within a special structure, the nucleus. It was first observed in 1831 by the English surgeon and botanist Robert Brown. (His name is perpetuated not as the discoverer of this cardinal component in the cell's anatomy but as the one who discovered Brownian movement, the random, jerky, zigzag motion of colloidal particles as they are buffeted by smaller molecules of high speed.) The nucleus is a spherical or ovoid structure encased within a membrane, which can be seen under the great magnification of the electron microscope in Figure 2.9. Usually there is only one nucleus within a cell; however, some cells are endowed with two or more. On the other hand, not every cell possesses a well-defined nucleus. For example, whether bacterial cells contain a genuine nucleus was a controversial topic among cytologists until recently. The electron microscope

DR. MAX F. PERUTZ

Fig. 2.8. Model of the hemoglobin molecule

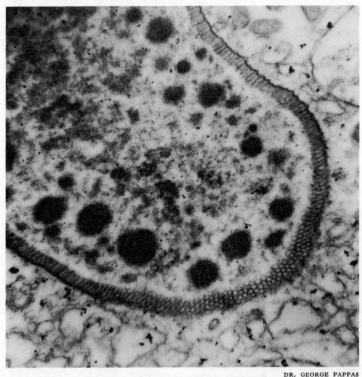

Fig. 2.9. The membrane which girds the nucleus

resolved the argument and it is now generally conceded
that bacteria are devoid of this structure. (The bacteria do,
however, have the usual genetic apparatus, although it is not
packaged in a nucleus.)

During part of its lifetime the nucleus appears to be a homogeneous structure except for a small sphere within the sphere, the
nucleolus, which stands out as a discrete area. The precise role
of the nucleolus in the cell is not well understood at the present
time. The nucleus sometimes gives the impression of transient,
unchanging serenity. This period in the life of the cell was

therefore called by early cytologists the "resting" stage. The description is valid only so far as visible changes are concerned. But beyond the reach of even the most powerful electron microscope there is a maelstrom of activity going on within a resting cell, with explosive speed and boundless variety. The most elaborate factory with the longest assembly line is a toy compared to the cell's machinery during the "resting" stage. Every component of a living cell is being manufactured at this time at a sufficient rate to ensure that when the time for division arrives there is a full stock to draw on to shape the many structures needed by the two daughter cells. During the resting state—which is also called the interphase—the chromosomes are diffuse and give the appearance of aimless disorganization. However, the appearance is misleading, for analysis reveals that the chromosomal material is increasing to ensure that there is an adequate dowry of this precious substance for each of the daughter cells. The first panel in Figure 2.10 is a simplified image of a cell in this resting state. At some point during the accumulation of the cell's substance, stock is taken, a decision to divide is reached, and the first steps to carry out the decision begin. Two pairs of minute structures—the centrioles—which lie just outside the nucleus begin to move apart, and they stretch between them gossamer threads or spindles (Panel 2). By the time the centrioles reach the opposite ends of the cell the chromosomes become more coiled and condensed. This is the beginning of the so-called prophase (Panel 3). Then the membrane, which up to now had encased the nucleus, begins to crack and crumbles like the walls of Jericho. The potent molecular trumpets in this case must be some enzymes, released perhaps from some lysosome. The thin threads of the spindle can now reach across the whole cell unhindered by the barrier of the nuclear membrane. By some mechanism unknown to us at the present time the threads of the spindle become enmeshed with small bodies within the chromosome. (Their position in the approximate center of the

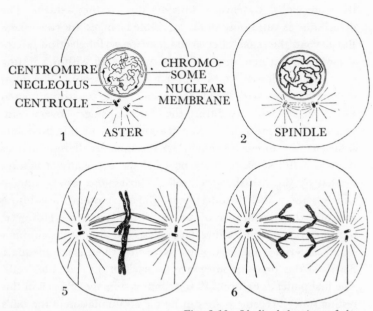

CENTROMERE
NECLEOLUS
CENTRIOLE

CHROMO-
SOME
NUCLEAR
MEMBRANE

1   ASTER

2   SPINDLE

5

6

Fig. 2.10. *Idealized drawings of the*

chromosome gives them their name, centromeres.) Whether the entanglement of the centromere and spindle thread is achieved by some molecular hook reaching out or whether the centromeres themselves spool out their threads to meet the spindles is unknown. Whatever the lassoing mechanism, the chromosomes are now secured to the centrioles like an incoming ship tied by a tossed mooring line to a post on the wharf (Panel 4). The chromosomes, which are now well-defined, rod-shaped structures, are dragged into the "equator" of the cell and are lined up, equally divided, like sausages carefully doled out by a judicious parent between two equally deserving children. This is the so-called metaphase, which can be seen in an idealized image in Panel 5. A movement toward the opposite poles slowly begins. And now the profound wisdom underlying all of these convolutions and alignments becomes obvious. The chromosome is a long, frizzly coil. When

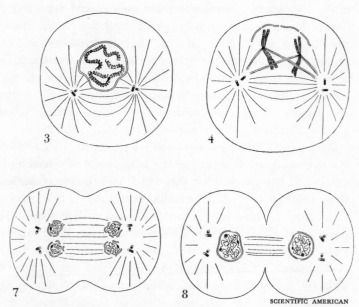

events occurring during mitotic cell division

fully extended the threads of the human chromosomes could be
a meter long. During the prophase stage of mitosis this tangled
thread is wound into a twin set of forty-six chromosomes, each but
a few millionths of a meter long. Like so much fuzzy unmanage-
able wool, the stuff is carded and wound into small tight spools.
Then during the movement of the anaphase (Panel 6) the
chromosomes can break cleanly without danger of entanglements
of the coils, assuring a clean separation and even distribution of
the chromosomal material. The dragging of the chromosomes
now can begin. That it is really a drag, the spindle fiber being
taken up like a cable by a capstan, is obvious from the shapes the
chromosomes assume during their voyage. They look like a
piece of rope snared by a fisherman's hook: The point of attachment swims ahead of the rest. Depending upon the location of
the attachment of the spindle fiber—in other words the position

of the centromere—the chromosomes often assume characteristic shapes during their migration toward the poles. Some assume the shape of a "v" others of a "j," but this hydrodynamic distortion is by no means the only influence which determines the shapes of chromosomes. The distance traveled by the chromosomes can be considerable—as much as five or ten times their own length. Their speed, however, is less than reckless—about a millionth of a meter in a minute.

At the end of their journey—the telophase, panels 7 and 8—the chromosomes, as if tired of the long period of confinement, uncoil and stretch. But lest they reach into the cytoplasm and thus wreak havoc with the cell's organization, new nuclear membranes are spun around them. The cell walls or membranes begin to be pinched off in the center and soon there are two young daughter cells complete in every way, ready to start a life of their own. If this mitotic division is repeated enough times to yield 10^{14} cells with proper differentiation so that the appropriate specialized cells are produced, then we have a whole new organism, a wondrous creature, such as the reader of this book.

From these overly simplified, schematic diagrams the reader may gather the false impression that cytology is child's play requiring little imagination or intellect. In fairness to cytologists, now let us see what is really visible under a microscope. In Figure 2.11 the various stages in mitosis in a white-fish embryo can be seen. It should be apparent that visual imagination in generous measure was needed by the early cytologists to deduce the intricate, sequential steps in mitosis from images such as these. And at that, this sequence in the cell division of the white fish, as in the case with all illustrations, is the best view obtainable. (The writer always had difficulty seeing what his professors of biology claimed *they* saw.)

One of the pragmatic tests a hypothesis must pass in order to gain validity is the prediction of the existence of new

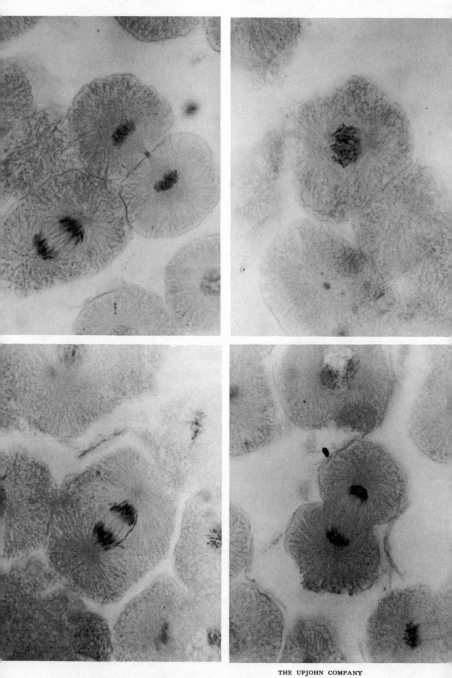

Fig. 2.11. Mitosis in the white fish embryo

phenomena. This can be done with high frequency in the physical sciences, for, inasmuch as God is a mathematician, there is a regularity, an order in the universe. But when it came to the creation of living organisms His mathematics became awfully involved. We are here, as we shall point out during the course of this book, as a result of an uncountable number of mutations. A living organism of today is an integrated product of all of those chance alterations. But in the midst of this welter of change some basic attributes of structure and patterns of function had to endure. No living creature that lost its nucleic acids or proteins survived, no organism exists without the battery of life, ATP. Even the bacteriophage, this simplest speck striving to be counted among the living, contains nucleic acid, protein, and ATP, although by itself it is unable to make any of these prime structures of life.

In living functions, too, there must be some mechanisms so fundamental no organism can survive their derangement. The orderly apportioning of the chromosomes into the sex cells is such a basic mechanism that it could have been predicted— provided one had had the insight of a genius. August Weismann (1834–1914), a German biologist, had such insight. He considered what might happen if each sex cell—the male and the female—contained the full complement of the chromosomes of the parent. In each fertilization, therefore in each generation, the chromosome number of the offspring would have to double. In not too many generations a cell would be as crowded with chromosomes as a football stadium and would have to be as big to accommodate all of them. Weismann, who in the belief that hereditary traits are entrusted to the germ plasm rejected the theory of inheritance of acquired characteristics, could not countenance such chaos in the cell. He made one of the few bold predictions in the history of biology. The number of chromosomes must be reduced to one half in every sex cell.

Patient work by several biologists in the last decade of the nineteenth century proved the soundness of Weismann's predictions. During the shaping of the germinal cells, both male and female, a reduction in chromosomes to half the total number in the parent's body cells occurs. This kind of cell division is called meiosis—from the Greek "to reduce." The diminution of chromosomes to one half in the "haploid" sex cells can be seen under a microscope during meiosis, and, after chemical analysis of the stuff of the chromosome became perfected, it could be demonstrated by chemical quantitation as well.

The mechanism within the cell which accounts for Mendel's laws now became clear. Let us return to his experiment with the red and white peas. Mendel concluded that the potential color locked within the seeds was controlled by a "factor" transmitted to the offspring from the parents through their sex cells. At the suggestion of the Danish geneticist Johannsen we call this factor a gene. Apparently two complementary genes are present in the body cells of an organism to achieve a trait such as color in the sweet pea. The *appearance* of the pea is called its phenotype; the permutation of the genes within its cells is called the organism's genotype. In a pure-red pea the phenotype is of course red and the genotype is red · red. In a pure white pea the phenotype is white and the genotype is white · white. During meiosis each of these peas will produce a sex cell containing one gene for this trait. The offspring from such a cross-fertilization will have a genotype of red · white. But its phenotype will be red. In other words, red is dominant over white. What the molecular mechanism of dominance is we do not know to this day. This offspring— it is called the F_1 generation from "filial"—is a hybrid, for its genes for the color factor are not homogeneous. When this F_1 hybrid produces sex cells meiosis will ensure that the complementary genes, red and white, are evenly distributed. When these sex cells fertilize each other the resulting plants of the F_2 generation will have the following probability for their genotype: There will be one red · red, two red · whites, and one white · white.

The phenotype will be three reds to one white because any plant containing a red gene will be red. This then is the cellular mechanism that is the basis of Mendel's patiently gathered statistics. The reader can see the same mechanism presented diagrammatically in Figure 2.12. The genotype is represented by the color of the misshapen sausages which represent a gene, the phenotype is expressed by the shading of the whole circle.

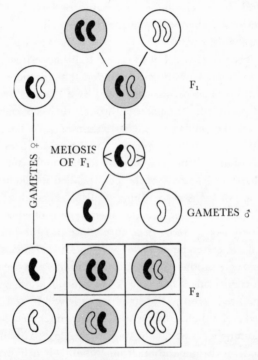

Fig. 2.12. Diagrammatic representation of color inheritance in the sweet pea

From the correlation of the work of cytologists and of geneticists, the general pattern for the transmission of hereditary traits was evident by the second decade of this century. As the cell divides so does the precious chromosome, thus ensuring the continuity of the functional and structural capacity acquired by the cell during its millennia of history. An understanding of the molecular mechanisms by means of which the chromosome achieves its purpose has emerged just recently. The explanation of our current view of those molecular mechanisms will be the burden of the rest of this book.

The first living cell that stumbled upon the method of cell division achieved immortality, for, if we assume that this wondrous ability was discovered only once, then the trillions of cells alive today are all descendants of that inventive ancestral cell.

3 THE ARCHITECTURE OF NUCLEIC ACIDS

I. THE BUILDING BLOCKS

THE TROUBLE with chemistry is the people who teach it. Instead of introducing the student to the beautiful logical sequence which is one of the major triumphs of the human mind, most teachers pour in front of their students a hodgepodge of trivia to be committed to memory. The hapless beginner is so mesmerized by detail he is lost. He cannot see the majesty of the structure for the molecules.

The source of the trouble is, of course, that the teacher himself probably failed to savor the joy of the logic of the discipline he is teaching. For the sad truth is that the introduction to chemistry in our country is performed at the high school level by anyone the principal can press into service and in most universities by impatient graduate students eager to get back to their laboratories and the work on their doctoral dissertations. The guidance of the beginner should be entrusted to individuals who have a mature knowledge of the science and of the structure of the logic on which it was built. Fluency in expression does no harm either.

In the next twenty pages I will try to condense enough of the logic of chemistry to enable the reader to follow our discussions of chemical structures in this book.

The modern view of the makeup of the elements which abound on the surface of our earth was first stated by the English teacher John Dalton (1766–1844), who proposed the atomic theory of matter. (To be sure, such a hypothesis was not original with Dalton. The Greek philosopher Democritus [460–370 B.C.] held essentially the same view. But this is the history of almost any pervading idea. Some Greek or Chinese or, in recent days, some Russian, had proposed it first.)

Dalton was a schoolteacher who in his leisure dabbled in science. He was thus doubly a pioneer. He was one of the earliest of scientists who bought for himself the freedom for the pursuit of his studies by teaching. The alchemists were supported by the largesse of kings and princes and they paid a heavy price for their support. They were expected to do their master's bidding and engage in pursuits the kings could understand— change base metals into gold, search for the elixir of life.

Needless to say, the atmosphere was less than conducive to any real creative effort. Any project a moronic king or prince could grasp was hardly worth pursuing. Nevertheless, more by accident than design, these alchemists did stumble on precious bits of knowledge. Thus Hennig Brandt, a lackey of some German prince, in 1669 cooked together sand and urine. He was undoubtedly motivated for this hanky-panky by dreams of extracting something mysterious from this biological waste product. As it happened, he was successful beyond his dreams. He discovered the element phosphorus. The product was indeed mysterious: It glowed in the dark! The mechanism of the glow—phosphorus combines with atmospheric oxygen and some of the energy of this reaction is liberated as light—remained obscure for over a hundred years, until Lavoisier demonstrated the real nature of combustion.

Lavoisier, while not typical of the eighteenth-century scientist, was already free of the shackles—both intellectual and financial—imposed upon the alchemist. He was a gentleman scientist. He financed his researches by other gainful pursuits. He was a farmer-general—a euphemism for tax collector—to Louis XVI, and was thus able to afford what passed in those days for a scientific laboratory. There, assisted by his charming and devoted wife, he performed the searching experiments that served to usher in the modern era of chemistry. The Alsatian chemist Charles Wurtz was to begin a textbook of chemistry a few generations later with the almost justified affirmation: "Chemistry is a French science. It was founded by Lavoisier, of immortal memory."

Lavoisier, along with another French chemist, Joseph Proust, laid the quantitative foundation on which Dalton could erect his atomic theory.

The alchemists had a penchant for endowing chemicals with mystical anthropomorphic attributes to explain chemical activities. When hard pressed for an explanation they would invent imaginative expedients. For example, the alchemist G. E. Stahl concocted a charming scheme for the mechanism of combustion. When a substance burns, said Stahl, "phlogiston" escapes from it. The residual ash, therefore, weighs less. When, later, it was discovered that sometimes the ash *gains* in weight the phlogiston theory was undermined—but only temporarily. Phlogiston, it was decided, was a versatile entity which could have either positive or negative weight.

Lavoisier designed a flawless experiment to probe the phlogiston theory. He heated a weighed amount of mercury in air in a sealed system so that he could measure both the gain in weight of the heated metal and the diminution of the "salubrious portion of the atmosphere," in other words, of oxygen. In turn, he took known amounts of mercuric oxide—for that is the product of the reaction of mercury and oxygen—and heated it to a sufficiently high temperature to decompose it. He discovered from painstaking measurements that the oxygen thus released is exactly the amount which had originally combined with the mercury. And that was the end of the phlogiston theory. Lavoisier and his wife gave the phlogiston theory a decent burial. In a mock ceremony Mme. Lavoisier, dressed as some high priestess, consigned the phlogiston theory to a funeral pyre while their assembled friends chanted a solemn dirge. Science in those days was a fun profession.[1]

From the work of Lavoisier and of Proust it became evident that elements do not react with each other randomly. They are not like so much shapeless putty mixed with, say, oil in any ratio. Rather, elements react in definite proportions. Table salt, whether obtained by evaporation of sea water or from a salt mine or by burning sodium in chlorine, will always contain 39.6 percent sodium and 60.4 percent chlorine. The relative amounts of the starting materials do not affect the outcome. In other words one can start with 39.6 parts of sodium and 60.4 parts of chlorine and get table salt. If we start with a millionfold excess of either of the elements we still get salt, and the element in abundance simply remains as a surplus. To Dalton this could

[1] This happy and fruitful marriage was ruthlessly severed by a revolutionary tribunal. Lavoisier was guillotined in 1794 by order of the National Convention. Eleven years later Mme. Lavoisier married Count Rumford, the Massachusetts-born American renegade, minor scientist and toady to kings. This marriage turned out to be a disaster and ended in separation in three years. Count Rumford is reputed to have said that the guillotine would have been preferable. But from his own accounts he himself was not blameless. Mme. Lavoisier's wealth was not without its persuasive influence in enticing the gallant count into matrimony. However, once they were married Mme. Lavoisier-Rumford concentrated her lavish hospitality on the guests who attended her tastefully arranged soirees. Between soirees the count was fed the leftovers from the festivities. This assault on the count's palate and vanity prompted him on one occasion to lock out Madame's guests, and he hid the gate key so that the embarrassed hostess had to make her apologies over the fence. She retaliated by pouring boiling water on the count's favorite experimental flower beds.

mean only one thing. He reasoned that this could occur only if the atoms of an element were discrete particles, uniform in size and homogeneous in chemical propensities. This, then, was Dalton's atomic theory, the validity of which has been upheld right up to today. One assumption of Dalton's needed minor modification. He postulated that the atom was indivisible, and of course cleavage of the atom is today a routine exercise. But even in this instance Dalton can be considered strictly correct, for when an atom is split it loses its identity. It no longer is an atom of that particular element.

Dalton's hypothesis on the discontinuous, homogeneous nature of atoms could give no clue about their relative size or about their mode of combination. Do they have wide-ranging sizes and weights? Do they combine in the same ratios: one for one, or are some more polygamous than others? These were questions which could not even be approached in the dawn of the nineteenth century. But in 1811 Count Amadeo Avogadro, who later became professor of physics in Turin, had one of those illuminating visions of genius which points the way for generations to follow. Count Avogadro was pondering some experiments reported by a Frenchman, Joseph Louis Gay-Lussac. The latter, a professor at the University of Paris, reported in 1808 some observations he had made on the manner in which gases enter into chemical combination with each other. He found that when gases interact they do so in volumes of small whole numbers, and, if the product is also a gas, it too will have a volume which is a small whole number. In other words, if we take one volume of gas A it will react with exactly one volume of gas B and there will be neither A nor B left. Avogadro reasoned as follows: If Dalton is right and atoms of elements are discrete particles and they combine with each other as Gay-Lussac said, whole volumes for whole volumes, it must mean that the one volume of gas A must contain the same number of molecules as the one volume of gas B. In other words, equal volumes of gases contain equal numbers of molecules. The far-reaching significance of his hypothesis was lost on Avogadro. It was only his successors who seized his inspiration and applied it to the weighing of the atom and thus started us on the path toward the current knowledge of atomic structure.

The stratagem for determining atomic weights based on Avogadro's hypothesis was as follows: We cannot determine the weight of a single atom. It is too elusive for capture and too small to weigh on the most sensitive of balances. But since equal volumes of gases contain the same number of molecules, we could weigh the same large volume of two different gaseous elements and obtain the relative weights of the atoms. To make this clear let us look at an analogy.

Suppose we wished to obtain the weight of a poppy seed and of a pea and the only balance available for the task was a roadside scale on which trucks are weighed. We could obtain a *relative* weight of the two seeds by loading a huge number, say a million, of each on the crude scales.

Similarly we could obtain the *relative* weight of hydrogen and oxygen by weighing, say, ten liters of each. The former will weigh about 0.9 grams and the latter will weigh 14.4 grams, or sixteen times as much. In other words, the oxygen atom must be sixteen times as heavy as the hydrogen.

This method of determining relative atomic weights occurred to Stanislao Cannizzaro, a professor at Rome, and to Lothar Meyer, a professor at Tübingen. Obscure though their names are to the general public, they must be counted along with Avogadro among the most brilliant minds of the nineteenth century. Their inspiration started us on the road to the understanding of the structure of the atom, the road which led relentlessly and inevitably to Stagg Field in Chicago, where the knowledge gained during the hundred years after the weighing of the atom enabled Fermi and his co-workers to release the energy locked within the atom.

Soon after Cannizzaro and Meyer showed the way, other chemists took up the task and determined the relative weights of the various gaseous elements. Once we had relative weights of gases, the atomic weights of some of the metals could be obtained. The approach here was different.

Let us take the case of silver. Since it is a solid its atomic weight could not even be approximated. However, silver enters into a chemical combination with a gas whose atomic weight was known, chlorine. The product which is formed contains 75 percent silver and 25 percent chlorine. Now, if we make the assumption that the two elements combined with each other in the simplest ratio of one atom of silver with one atom of chlorine then the atomic weight of silver can be calculated from the known atomic weight of chlorine. Since the product, silver chloride, contains three times as much silver as chlorine the atomic weight of silver must be three times as large. Since the atomic weight of chlorine is 35, silver must be $3 \times 35 = 105$. (It is actually 107.)

The total information described above can be put simply in chemical shorthand as follows:

Silver	Chlorine	yields	Silver Chloride
Ag^+	+ Cl^-	→	AgCl
107	35		142

(The plus and minus signs above the silver and chlorine did not come in until some forty years later.)[1]

---

[1] For the sake of brevity and clarity, the section above is oversimplified. Actually much hard work and keen insights were needed to distinguish between atomic weights and molecular weights. Gaseous elements, for example, do not exist as simple atoms but rather as tightly bound pairs of them. To complicate the task of the early chemists further, elements do not always react with each other in a simple one to one ratio; an atom of aluminum although but one fourth the atomic weight of silver combines with three atoms of chlorine. If the reader is interested in the resolution of these once baffling problems he will have to turn to a more extensive presentation of the history of chemistry.

The patient weighing of the atoms by many chemists yielded by 1865 the atomic weights of sixty-two different elements. Certain similarities in behavior among some of the elements had already been correlated. Some elements are gases, others are solids, some have certain metallic attributes, others are non-metals. Some form colored compounds when they react with, say, chlorine, others form white solids. The temperatures at which the solid metals melt is wide-ranging. All of these attributes or properties of the elements were carefully catalogued and it was known that several elements share similar attributes.

The gathering of information in science is a steadily progressing process. The *understanding* of the gathered information, however, comes in jumps. In the jargon of science one could say: "The accumulation of information moves in a linearly ascending curve, but understanding is achieved in quantum jumps."

It is a frequently observed characteristic of this dual advance that at a point where a sufficient amount of information has been gathered several men of ability can make the jump to the higher level of insight and understanding. The seventh decade of the last century was such a period of sudden advance in our knowledge of the structure of the atom. Several men began to notice a certain periodic recurrence of similar attributes among the elements when they were lined up in order of increasing atomic weights. Beguryer de Chancourtois, in 1862, was the first to make such correlations. He said: "The properties of the substances are the properties of the numbers." In 1865 the Englishman John Newlands detected this repetitive regularity in every eighth element and called his discovery the "law of octaves." He emphasized the importance of 8 both in music and among the elements. This kind of numerological hanky-panky would have created a stir among the alchemists, but the scientists of the second half of the nineteenth century had exchanged mysticism for rational materialism and rejected Newlands' musical elements.

The same periodicity of 8 stated merely as a Law of Nature was readily accepted four years later. The new law was formulated simultaneously by two men of rare insight: Lothar Meyer and Dmitri Ivanovich Mendeleyev, a professor at the University of St. Petersburg. The latter developed his discovery with boldness and imagination, and the Periodic Law, as it is called, is associated with his name. Mendeleyev was writing a textbook of chemistry and decided to organize the known elements in order of their increasing atomic weights. As happens to authors of books, they know more about their subject during the period of writing than ever before—or ever after. From Mendeleyev's writing labors a pattern emerged. He said that he saw suddenly at night, in his sleep, the periodicity with which elements of similar properties appear if they are arranged in order of increasing atomic weights. He wrote: "The elements, arranged according to the order of their atomic weights, represent a clearly recognizable periodicity of the properties. The magnitude of the atomic weight determines the character of the elements."

With the lighter elements Mendeleyev's pattern was perfect: Every eighth element was a heavier counterpart the one an octave earlier. But he ran into trouble with some of the heavier ones. Zinc and arsenic followed each other in the sequence of increasing atomic weights. They differ by ten atomic weight units and usually the atomic weights of consecutive elements differed by only three units. Moreover, the periodicity of attributes did not match if arsenic was placed right next to zinc, but if two gaps were left and arsenic was put three units away everything fell into line. Mendeleyev now demonstrated that there was more to him than the capacity for a single flash of insight. He apparently had enormous faith that there is an order in Nature and he boldly stated that the double gap between zinc and arsenic represented two missing elements which, however, must exist. He went further than that. He predicted many of the properties of the unknown elements, being confident that they would follow the periodicity of properties expected of them.

The uncanny accuracy of his prediction can be seen from the example below:

Prediction by Mendeleyev in 1871		*Found by Winkler 1886*
Eka-silicon		Germanium
Atomic weight	72	72.3
Specific gravity	5.5	5.47
Atomic volume	13	13.2
Color	Dirty gray	Grayish white

It should form a compound with chlorine, whose formula is ECl_4,
which should be a liquid — $GeCl_4$ is a liquid
with specific gravity 1.9 — Sp.g. $= 1.887$
and boiling point below 100°. — B.p. $= 86°$

Mendeleyev predicted the existence and properties of three new elements and in every case his predictions were borne out as well as in the case of germanium. Seldom in the history of science has there been such a brilliant demonstration of the capacity of the human mind to pierce the complexity of our universe and then to weave a pattern and predict with confidence new facts which must fit that pattern. The collective achievement of Avogadro, Cannizzaro, Meyer and Mendeleyev should be classed along with those of Kepler, Newton, or Einstein. However, in the case of the latter group absolute confirmation crowned only Einstein's theory—of the convertability of matter and energy—during his lifetime. Kepler's hypothesis received unequivocal proof only recently when the first Sputnik achieved unfettered orbital motion. Newton's hypothesis of the gravitational interdependence of all heavenly bodies was stunningly confirmed in 1846 when two astronomers, Adams of England and Leverrier of France, independently calculated from the perturbations of the orbits of the known planets the existence of another one. The German astronomer Galle then pointed his telescope to the spot where the calculations guided him, and there was the planet Neptune.

The tempo of the acquisition of knowledge of the atom accelerated after Mendeleyev. The French physicist Henri Becquerel made one of those classic discoveries where a false hypothesis leads to lasting truth. He discovered that certain minerals give off piercing rays which, like X rays, can penetrate through opaque barriers. Becquerel's colleague Pierre Curie and his wife, Marie, started a search for the most potent source of these radiations. With the confidence gained from Mendeleyev's confirmed prediction of the existence of elements yet unknown, the Curies concentrated the source of the puissant rays and isolated a new element, radium. Radium turned out to be the Rosetta Stone for deciphering the structure of the atom. Its relentless disintegration into electrical particles revealed the nature of the brick and mortar from which atoms are shaped. They are electrically charged particles held together in an atom by forces which are convertible to electro magnetic energy.

Three different emanations from the exploding radium atom were recognized. The alpha particle is a positively charged mass which is four times as heavy as the hydrogen atom. The beta particle turned out to be a rapidly moving, negative speck of matter, which is about 1/1800 the mass of the hydrogen atom. It was recognized as an electron, an elementary particle which had already been known from studies of the passage of electric current through evacuated tubes. Finally, the gamma radiation proved to be an electro-magnetic burst of energy on the order of X rays.

The energy surging out of radium proved to be stupendous. Radium, which has an atomic weight of 225, disintegrates into smaller particles, including the metal lead, which has an atomic weight of 210. Accompanying this disintegration, about 130 calories of heat are released by a gram of radium in an hour.

This in itself is not an overwhelming amount of energy, but the radium can keep this up at a diminishing rate for over 2000 years, and thereby it is more than 300,000 times more efficient as a source of energy than a gram of coal. Hence, radium was the first known source of atomic energy.

But radium provided more than the image of the disintegrating atom; it provided a new powerful tool for the exploration of the inner structure of the atom. This tool was wielded effectively by Ernest Rutherford, a New Zealand born professor of physics at the University of Manchester. One of the particles which issues from the debris of radium is the alpha particle. As we said before, the alpha particle is positively charged and has a mass four times that of hydrogen. Actually the alpha particle is a charged helium atom. (The element helium was discovered in the gases of the sun from spectral lines which had no counterpart to those issuing from any earthling element. It was discovered by the English astronomer Lockyer, who named it very aptly: helium. Later, helium was found on earth as well.) This atomic missile issuing from the core of the radium atom goes about 5 to 6 inches in air before being halted by the collisions with the trillions of atoms in its path. Rutherford placed a thin sheet of gold on top of the source of alpha particles and waited to see what the presumably continuous barrier would do to the surging particles. To his great surprise they almost all went through: Only about one in 8000 met an obstacle that made that alpha particle bounce back. Rutherford was shooting poppy seeds through chicken wire! This simple experiment revealed the structure of the atom. It has a solid core—the nucleus—which is minute compared to the whole. This nucleus must be positively charged (it repels the positively charged alpha particle). It was already known that the atom contains negatively charged particles—or electrons. (All one has to do is heat a metal to incandescence and electrons keep jumping off like popcorn from a hot skillet.) Now, since the nucleus is positively charged, and the atom itself is neutral, there must be electrons outside of the nucleus. The study of the position of the extranuclear electrons has been one of the most intriguing, if somewhat esoteric, intellectual pastimes of the past six decades. Physicists have interpreted the different spectra which atoms emit to be the result of the distribution of the electrons at various distances from the nucleus. It was the Danish physicist Niels Bohr who first suggested that the atom is a microcosmic replica of our solar system. At the center is the nucleus around which the electrons whirl at different distances, with characteristic orbits like so many planets. As to the relative sizes and distances involved, someone has fashioned the following analogy: If we assume the nucleus of the atom to be the size of a football which is placed in the center of New York, then an electron is the size of a pea, the first of which will be whirling as far out as Philadelphia, the next line of these pea-electrons will be orbiting through Pittsburgh.

So much for the structure of the atom. But it is clear that so far we have not suggested how atoms interact to form compounds. A surprising discovery and a brilliant intuitive insight provided the answer for the mechanism of chemical union between the atoms of elements. In 1893 Lord Rayleigh, an English physicist, decided to determine accurately the density, or rather the atomic weight, of nitrogen. Being a meticulous scientist he prepared the gas from different sources, from the atmosphere and from chemical compounds of nitrogen. When he was finished with his measurements he found that the atomic weight of nitrogen from the atmosphere was 14.07 (oxygen having an arbitrary value of 16.000) and the nitrogen from the pure chemical compounds 14.005. Rayleigh was confident that the discrepancy of seven parts in 1400 was not due to an experimental error on his part and he published his findings. A year later the British chemist William Ramsy exposed the presumably pure nitrogen from the atmosphere to hot magnesium. The hot metal did not sop up all of the gas as expected. About 1 percent remained untouched, and nothing could be done to induce this residue to combine with any other chemical. This residue proved to contain several new, unknown elements, the so-called inert gases. English scientists of that era had sound classical education and the elements were furnished with attractive Greek names: argon (idle); neon (new); xenon (stranger); krypton (hidden).

The new elements offered a challenge to Mendeleyev's classification: Could they be fitted into the pattern of periodicity with the pre-existing elements? Their atomic weights were at such intervals that they fitted into the table perfectly, forming a new family of elements with a recurrent, new, periodic chemical property. The property was, however, unique: These elements did not react with any of the other elements.

Two imaginative men now entered the stage in the history of the development of chemical ideas.

The first one was a Swede, Svante Arrhenius, who while still a graduate student made a penetrating observation on the behavior of substances in solution. Some solutions conduct the electric current; some do not. Those substances which are electrical conductors behaved in solution as if they were separated into their component elements but with an electrical charge on them. Thus table salt—NaCl—behaves, in solution, according to Arrhenius, as if it contained Na^+ and Cl^- particles. These charged atoms Arrhenius named ions, from the Greek word to go. (The aptness of the name stems from the ability of ions to migrate in an electric field toward the oppositely charged pole.) Now the interesting fact about the formation of ions by the different elements is that the elements which come in the increasing order of atomic weights just before an inert element form negative ions, and those which come right after an inert element form positive ions. The elements in between tend to form compounds which are not ionic, and therefore their compounds do not conduct the electric current.

The chemist who first synthesized all of this information into a unified working hypothesis was Gilbert N. Lewis. He was one of a handful of highly gifted young Americans who at the turn of the last century were slowly converting the role of Americans from readers of science to leaders in it. Michaelson and Millikan in physics, Gomberg in organic chemistry, T. W. Richards in analytical chemistry, and Lewis in physical chemistry were American names to be reckoned with in science in the early part of this century. (Three of these five were Nobel Prize winners, but Lewis was not among them. However, the committees awarding those coveted prizes do blunder from time to time, albeit not as egregiously, or as often, as the committees which award the same prizes for literature.)

Lewis, who was professor of chemistry at the University of California, pondered the available information and came up in 1916 with a beautifully simple hypothesis which unified much of the apparently disparate facts about the reactions of the chemical elements. He reasoned as follows: Chemical union between atoms of two different elements must be achieved by the loss or gain of electrons. If an element loses an electron—which is a negative particle—the atom must by this deficiency become positively charged. In turn, an atom which is the recipient of one or more electrons during a chemical reaction becomes negatively charged by reason of the extra burden of electrons. The inert gases do not enter into chemical combinations.[2] In other words they neither gain nor lose electrons. Therefore, argued Lewis, they must have a stable system of electrons. The elements other than the inert gases must strive to achieve the stability of the inert gases by gaining or losing electrons and thus mimicking the structure of the nearest inert gas. If a given element has one, two, or three fewer electrons than the nearest

---

[2] Exceptions have been found recently. For example, xenon reacts with the gas fluorine to form XeF_6. However, it is an unstable compound, as if the xenon resented the intrusion of the six fluorine atoms into its electronic sphere.

inert gas, then during chemical union it will tend to capture one, two, or three electrons and become a negatively charged ion. If an element has one, two, or three electrons *more* than the nearest inert gas then that element will tend to strip off these electrons during a chemical union and become a positively charged ion. The periodicity of chemical properties which Mendeleyev noted is a function of the number of electrons an element gains or loses in a chemical reaction. For example, all those that *lose* one electron are the alkali metals, the most familiar of which are sodium and potassium; those that *gain* one electron are the halogens, chlorine, bromine, and iodine.

The elements which are midway between two inert gases, in other words, those that have four electrons in excess of the inert gas, are unique in their chemical combinations. They tend neither to gain nor to lose electrons; rather, they enter into partnerships with other elements and share their electrons to achieve mutually the stability of the inert gases. Since in this kind of chemical combination the atoms are not converted into autonomous ions, the elements remain attached to each other and can form large aggregates by the attachment of many atoms. These accretions can contain as few as two different species of elements or they can be mosaics of a half-dozen different ones; they can be as low in molecular weight as an oxygen atom (16) or they can be gigantic constellations reaching into the millions in molecular weight.

The element carbon is the lightest of the elements which can offer four electrons for chemical combination and it has the greatest propensity for entering into electron-sharing liaisons with other elements and also with other carbon atoms. Thus carbon is capable of forming fantastic conglomerates of vast size and almost infinite variety. The English astronomer Hoyle has suggested that life was created the moment matter was created. This may be true in the sense that the spark of life started flickering in the four valence electrons of carbon. Only from so versatile an element which has the ability to shape itself into an infinity of patterns, each with unique properties, could so wondrously complex a system as life arise.

The development of the chemistry of the carbon atom—organic chemistry as it is called—is one of the truly remarkable achievements of the human mind. The achievement is all the more remarkable because it was won with rather crude procedures and even cruder symbolism. However, the pictorial imagination and the logic which guided the chemists of the past century and a half were faultless.

With a little patience and interest the symbolic logic on which organic chemistry is built is easily followed. The simplest organic molecule known is marsh gas or methane. It contains only carbon and hydrogen, nothing else.

Since it is a gas, its molecular weight—compared to the oxygen atom—is easily determined: it is 16. The carbon atom is known to be 12 and a hydrogen atom to be 1. From this information alone we can deduce that the molecule must contain one carbon and four hydrogen atoms, i.e., we can write its formula as CH_4, or in two dimensions:

$$H-\overset{\displaystyle H}{\underset{\displaystyle H}{\overset{|}{\underset{|}{C}}}}-H$$

Methane

The line between the carbon and a hydrogen atom is a shorthand symbol indicating a bond formed by the sharing of two electrons between the two atoms. In other words a carbon and a hydrogen atom uncombined might be represented thus:

$$.\overset{.}{\underset{.}{C}}. \qquad \overset{\times}{H}$$

(The dots and cross represent the electrons of carbon and of hydrogen respectively.) After the formation of methane the five atoms now may be symbolized as follows:

$$
\begin{array}{ccc}
& \text{H} & \\
& \overset{\times\,\cdot}{} & \\
\text{H}^{\cdot}_{\times}\ \ \underset{\cdot\,\times}{\text{C}}\ \underset{\times}{\cdot}\ \text{H} & \quad\text{or}\quad & \text{H---C---H} \\
& \text{H} & \text{H}
\end{array}
$$

Note that there are eight electrons around the carbon atom, four of its own and four adopted from and shared with four hydrogen atoms. The carbon atom has thus acquired the eight-electron stable structure of the next higher inert gas, neon.

There are a variety of ways to make more complex organic molecules. For example, one or more of the hydrogen atoms in methane can be replaced by chlorine atoms in a reaction of the two gases methane and chlorine. If the methane is in excess then it is relatively easy to isolate a product in which only one hydrogen is replaced by chlorine. (This is merely the result of the operation of the laws or probability at the molecular level. If there is a large excess of methane molecules compared to those of chlorine, the probability of single substitutions is greater than that of multiple substitutions, simply because under the circumstances it is more probable for a chlorine atom to encounter a virgin methane molecule than a chlorinated one.) We symbolize a monochlorinated compound as:

$$
\begin{array}{c}
\text{H} \\
| \\
\text{H---C---Cl} \\
| \\
\text{H}
\end{array}
$$

If *chlorine* is in excess we may get products such as trichlor methane, which is chloroform,

$$
\begin{array}{c}
\text{Cl} \\
| \\
\text{H---C---Cl} \\
| \\
\text{Cl}
\end{array}
$$

or carbon tetrachloride

$$
\begin{array}{c}
\text{Cl} \\
| \\
\text{Cl}\!-\!\text{C}\!-\!\text{Cl} \\
| \\
\text{Cl}
\end{array}
$$

The beauty of the system of molecular weights based on Avogadro's hypothesis is that it works perfectly with these substituted compounds as well as with the original elements. Thus chloroform should have the additive weights of one carbon, one hydrogen, and three chlorine atoms. And indeed it does.

Nature abounds in vast numbers of organic molecules of wondrous variety. How do we determine their structure? Organic molecules can participate in a large variety of different reactions. To determine the structure of a naturally occurring substance the organic or biochemist wields these reactions as a painter wields his brush. For example, let us establish the structure of ethyl alcohol by synthesis. To do this we shall set out to make ethyl alcohol by a method which shuns the tedious harvest of the grape, pressing out the juice, and fermentation. If the methyl chloride (CH_3Cl) which we made is allowed to react with metallic sodium a reaction occurs producing sodium chloride and a new compound which contains two carbon and six hydrogen atoms.

Considering its source—two $\;\begin{array}{c}\text{H}\\|\\\text{H}\!-\!\text{C}\!-\!\\|\\\text{H}\end{array}\;$ groups—the new compound must have the following structure:

$$
\begin{array}{c}
\text{H} \quad \text{H} \\
| \quad\;\; | \\
\text{H}\!-\!\text{C}\!-\!\text{C}\!-\!\text{H} \\
| \quad\;\; | \\
\text{H} \quad \text{H}
\end{array}
$$

The chemist would symbolize the total reaction as follows:

$$
\begin{array}{ccccccc}
& H & & H & & H & H \\
& | & & | & & | & | \\
H\!-\!\!C\!-\!Cl & & Cl\!-\!\!C\!-\!H & \longrightarrow & H\!-\!\!C\!-\!\!C\!-\!H & +\ 2\ NaCl \\
& | & & | & & | & | \\
& H & & H & & H & H \\
Na & & Na & & & &
\end{array}
$$

The new compound, ethane, can also be exposed to chlorine and from the reaction a singly substituted chlorine product can be isolated. This compound, ethyl chloride, is symbolized as follows:

$$
\begin{array}{ccc}
H & H \\
| & | \\
H\!-\!\!C\!-\!\!C\!-\!Cl \\
| & | \\
H & H
\end{array}
$$

That the formula we assigned to it is a valid one is evidenced by its molecular weight: the sum of two carbons, five hydrogens, and one chlorine.

Now, if ethyl chloride is heated with water and sodium hydroxide (NaOH) the chlorine atom is exchanged for an oxygen bound to a hydrogen—OH. The new organic compound is

$$
\begin{array}{ccc}
H & H \\
| & | \\
H\!-\!\!C\!-\!\!C\!-\!OH \\
| & | \\
H & H
\end{array}
$$

If we take any source of naturally produced alcohol, be it rye, cognac, or vodka, and purify the alcohol by technically sophisticated distillations, we eventually obtain pure alcohol. The natural alcohol and the product made in the laboratory can now be put through a variety of tests and measurements which serve to fingerprint the products. The temperature at which they boil, the weight of a unit volume, alterations in the path of light as it passes through the liquid, molecular weight and percentage composition, these are the whorls in the fingerprints by means of which the chemist establishes the identity of the two liquids. And they turn out to be identical. Therefore we have established the chemical structure of a naturally occurring substance by the synthesis of it in the laboratory.

The term *synthetic* has acquired an opprobrious meaning indicating a poor substitute for the genuine. However, in the case of a biologically active substance, when the chemist synthesizes it he makes an identical duplicate of what nature has made. Vitamin C, or Vitamin B_1, or the hormone of the posterior lobe of the pituitary gland which induces labor in a pregnant female, any of these products is exactly the same whether made by the living cell or by the knowing hands of the chemist. It is impossible to tell the products apart by any chemical or biological tests.

Organic chemistry has changed our way of life. Medicine, nutrition, our clothing and dwelling all have gained in uncounted ways from the fruit of the logic and skill of the organic chemist. It is a profound source of pride to contemplate the products of this simple system of pictorial symbolism. But to those who really understand it the achievements of organic chemistry can even be the source of aesthetic joy. I recall that, when I first read Dr. Vincent du Vigneaud's brief description of his synthesis of the hormone of the posterior pituitary, I experienced the same spine-tingling excitement as when I first heard Toscanini conduct Beethoven's Ninth Symphony, or when I first read Melville's description of the Pequod sailing through the night, sails billowing and the fires under the melting kettles sparking the dark Pacific night. Lawrence Durrell wrote somewhere: "Science is the poetry of the intellect." Organic chemistry contains some of the noblest—and least appreciated—passages in the collected volumes of the poetry of the human mind.

Now let us return to the organic chemistry of the nucleic acids. Miescher had two students who carried on where their teacher left off at his untimely death of fifty-one.

One was R. Altmann, who kept purifying nuclein until it was free of contaminating proteins and who coined the term "nucleic acid" because nuclein, once freed of basic proteins, turned out to be an acidic substance. Another student of Miescher, Piccard, decided to search for the component building units of nucleic acids. By then organic chemists had identified about half a dozen different amino acids which are the building units of proteins. Piccard followed the pattern of search for the components of proteins; he cooked the nucleic acid with acid and he hunted for degradation products in the soup. He isolated two; one was an intact building block, the other was an artifact, a decomposition product of still another component. The substance which has stood up as a genuine component of all nucleic acids is guanine. Its structure was deduced by patient analysis and proved by actual synthesis.

Guanine

The isolation of guanine could give not even an inkling of the structure of nucleic acids. It served merely to indicate that the component units are different from those of the proteins.

The torch for the search for the structure of nucleic acids was next taken up by a student not of Miescher but of Hoppe-Seyler,

Albrecht Kossel. This continuity from teacher to student is characteristic in the growth of our knowledge in an area of science. Science can be mastered from books alone only at the most elementary level. For the journey to the frontiers of knowledge an experienced and willing master is needed as a guide. One of the most heart-warming traditions of science is the spontaneous generosity of many masters of our craft who guide fledgling scientists into their own areas of specialization and urge them to go on to new areas or to compete with their teacher, as they will.

And, of course, it will often come to pass that the pupil outshines the master in his total achievement. Hoppe-Seyler had such a student in Albrecht Kossel, who taught at the University of Marburg and at Heidelberg. Kossel was one of the flowers of German scholarship which was permitted to flourish in the wholesome atmosphere of the German universities at the turn of the century. Kossel and another giant of intellect, Emil Fischer, dominated the field of the chemistry of natural products. Scores of students from all over the world flocked to the laboratories of these men for pre- and postdoctoral training and thus their influence spread throughout the world. A great many middle-aged scientists of today all over the world are the second-generation crop of this once great school of German science.

In essence what Kossel and Fischer and their schools achieved was the identification of the building units of the large molecules which are the hallmarks of the living cell: the proteins, the starches, the complex fats, and the nucleic acids. The chemistry of the period was not up to the task of studying the integrated, large molecules. Chemists, therefore, addressed themselves to smaller tasks. They dismembered the complex products of the living cell and they tried to isolate and then identify the components in the debris. It is interesting that the functions of a large molecule were not recognized until the third decade of this century when the biocatalysts of the cell, the enzymes, were shown to be proteins.

The functions of the nucleic acids could not even be guessed at until the fifth decade of this century. The most educated guess was that the nucleic acids provided physical struts to support the physiologically potent proteins. Evidence for the neglect of the study of nucleic acids can be gleaned from a glance at the lecture schedules in a course in biochemistry thirty years ago. As much time was devoted to the study of urine as to the nucleic acids.

But despite this neglect, knowledge of the component units of nucleic acids kept increasing. The small number of investigators were compensated for by the energy and devotion of those who were digging in the field of nucleic acids. The most prodigious of these in the early part of this century was the Russian-born chemist Phoebus A. Levene, from whose laboratories at the Rockefeller Institute a veritable torrent of communications on nucleic acid structure poured forth. While most of his observations were sound, his interpretations left much to be desired. In contemplating P. A. Levene's work one is reminded of the apology of Blaise Pascal for the excessive length of one of his letters: "I had no time to write a short one." It would be wasteful to document in detail the meandering path of research in the maze of nucleic acid chemistry. Such detailed knowledge is useful only for a student of science. He should become steeped in the methods, ideas—both fruitful and barren—of the past so he can face the awesome task of contributing to the future. But my reader is only a visitor in the garden of science, so let me just show with pride the current blooms.

By 1944, which is, as we shall see later, a milestone in nucleic acid research, the following was known about nucleic acid chemistry. Nucleic acids are huge molecules which on degradation by acid or alkali yield nitrogenous bases, sugars, and phosphoric acid. Two different kinds of sugars were identified from nucleic acids from different sources: ribose and deoxyribose. These are compounds containing five carbon atoms festooned with hydrogen and oxygen atoms in a ratio of two hydrogens and one oxygen for each carbon atom.

The difference in the two sugars is the absence of an oxygen from deoxyribose. The two different sugars conferred their names on the nucleic acids from which they stem. Thus the nucleic acids containing ribose are called ribonucleic acid (RNA) and the other deoxyribonucleic acid (DNA).

Of the nitrogenous bases three are present in both RNA and DNA.

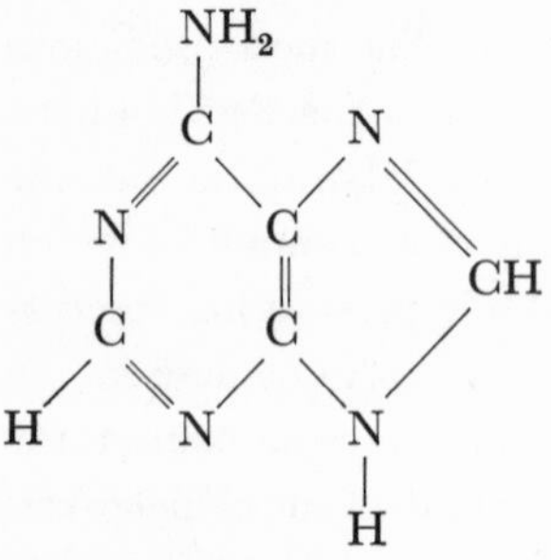

Adenine

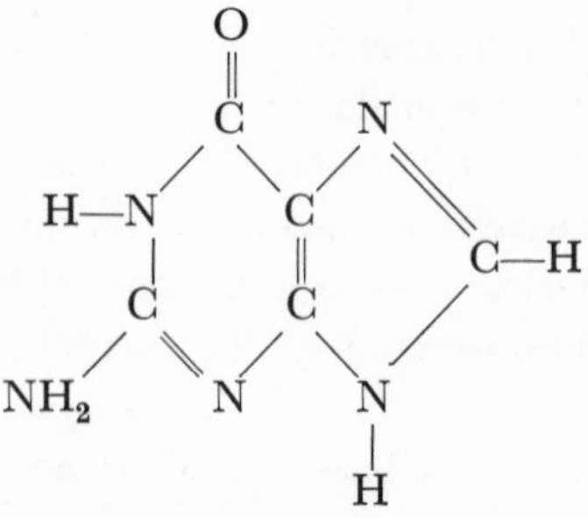

Guanine

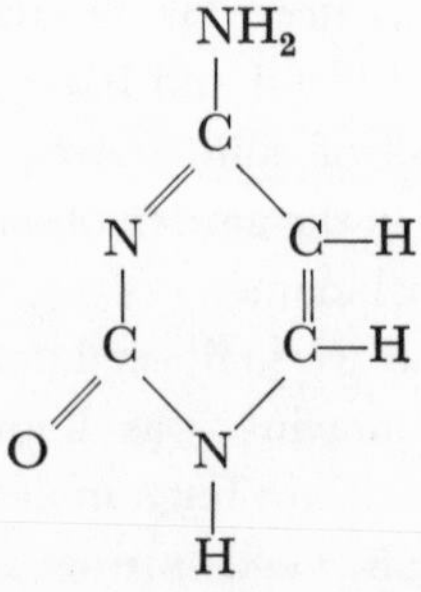

Cytosine

In addition to these three, RNA contains uracil.

$$\begin{array}{c}
O \\
\parallel \\
C \\
\diagup \quad \diagdown \\
H-N \qquad C-H \\
\mid \qquad \parallel \\
C \qquad C-H \\
\diagup \quad \diagdown \diagup \\
O \qquad N \\
\mid \\
H
\end{array}$$

Uracil

DNA on the other hand contains thymine, which is identical to

uracil except for an extra
$$\begin{array}{c}
H \\
\mid \\
-C-H \\
\mid \\
H
\end{array}$$
or methyl group.

$$\begin{array}{c}
O \\
\parallel \\
C \\
\diagup \quad \diagdown \\
H-N \qquad C-CH_3 \\
\mid \qquad \parallel \\
C \qquad C-H \\
\diagup \quad \diagdown \diagup \\
O \qquad N \\
\mid \\
H
\end{array}$$

Thymine

The reader might justifiably recoil from the contemplation of these arcane structures. I must remind him, however, that the biochemist has no choice over the products he studies. We must master the chemistry of a vast number of molecules of sometimes bizarre complexity, for life arose out of those compounds and indeed life continues as an interaction among them. Some of these simpler compounds have been around on the surface of the earth for billions of years. There is something extraordinarily stable in the configuration of atoms which produce, say, adenine. Let us see it again.

Adenine

It is easily seen that the compound contains carbon and nitrogen atoms in a one to one ratio. In recent years, as the successful victory over the laws of gravity has turned our gaze skyward and we begin to think anew of the possibility of life on cooled planets other than on our own cooling speck of cinder, the origin not only of life but of the compounds of life has come into the focus of attention. It was found that the building blocks of the proteins, the amino acids, are formed readily under conditions which had presumably existed on the cooling, primitive Earth. If electrical sparking—which can be considered the counterpart of electrical storms—is continued in a mixture of gases consisting of ammonia, water, and methane—then after a sufficient length of time amino acids can be shown to be accumulating under the influence of the sparking in this putative facsimile of the primeval atmosphere of our planet.

But the search for nucleic acid precursors under these conditions brought no rewards. However, a young man in Texas, Dr. Oro, had a bright idea. He decided to heat the simplest compound in which carbon and nitrogen have the one to one ratio of adenine. This is the cyanide ion CN^- and so he heated ammonium cyanide, NH_4CN. He extracted the melted mass and —lo and behold!—he could isolate adenine from the solution.

Why was this important member of the quartet of bases which make nucleic acids formed in the melted cyanide? Adenine was formed undoubtedly for the same reason that it had been formed and accumulated in the primeval seas: It is an extraordinarily stable compound. The fact is that there are probably literally hundreds of different compounds of carbon, nitrogen, and hydrogen formed as the high temperature excites the component atoms of ammonium cyanide. These new compounds are produced by the collision and cohesion of the activated fragments. But most of these compounds are too unstable to exist for more than fractions of a second. They therefore break up and are free for encounters that may yield more stable combinations. Adenine is such a molecular Rock of Gibraltar. (It withstands with impunity cooking with concentrated acid.) Therefore, adenine will accumulate at the expense of unstable transient structures. The compounds which abounded in the primeval seas on our cooling planet and were therefore available to serve as the building units for the edifice of life were thus selected for their roles by a process of atomic evolution. The governing principle of this evolution might be said to be: "The survival of the stablest."

Next came molecular evolution, the alignment into complex macromolecular aggregates of the stable compounds that had accumulated in the primeval seas. This was achieved with the aid of the most abundant and ubiquitous reagent: water. We retain to this date as a legacy from our primitive aquatic cradle not only the building stones of life but the mechanism of their assembly as well. The complex macromolecules of a contemporary living cell, the proteins, the fats, the starches, and the nucleic acids, are all pieced together by the elimination of one molecule of water from every two component units. From one unit a hydrogen(H^-) and from the other a hydroxyl(HO^-) group is detached. At the sites from which these groups were shorn the two molecules can now fuse, forming a double unit, or a dimer. But such a dimer can also undergo a similar accretive fusion and that is how the nucleic acids and the proteins, substances of infinite complexity and wondrous potency, are built by that fabulous architect: a living cell.

Now let us see how a nucleic acid might be formed by such fusion reactions. The simplest repeating building unit of a nucleic acid is a nucleotide, which contains a base, a sugar and phosphoric acid.

Here is the formula of such a unit containing adenine and ribose. (Therefore this would be a nucleotide from RNA.)

Adenylic acid

This nucleotide was formed by the elimination of two molecules of water, one to form the bond between the nitrogen 9 of adenine and carbon 1 of ribose, the second to bind the oxygen on carbon 5 of the ribose to the phosphoric acid. The writing of such a detailed formula is time consuming for the writer, baffling for the reader, and expensive for the publisher, so let us represent adenine by A and another base, say uracil, by U, and let us write the structure of a dinucleotide containing those bases.

A dinucleotide containing adenine and uracil

This dinucleotide had been formed by the removal of H_2O from the phosphoric acid and carbon 3 of the ribose attached to uracil. (It should be pointed out that the assignment of the bridge between carbon 5 and carbon 3 is not arbitrary. There is very sound evidence for that particular position and not, say 5 and 2. I shall present the evidence in Chapter 7.) But carbon 5 of the ribose, attached to uracil is free to receive—by the elimination of water—another phosphoric acid which, in turn could be used for hooking onto still another nucleotide, and so the accretion could continue almost *ad infinitum*.

In a very abbreviated shorthand this could be symbolized as follows:

(A = adenine, U = uracil, G = guanine, C = cytosine)

Or with even more abbreviated shorthand it would be:

Ap Up Gp Cp

Up until our arbitrary milestone of the discovery of the function of DNA in 1944, it was believed that nucleic acids were formed simply by the repeated fusion of four nucleotides in equivalent amounts. This so-called tetranucleotide hypothesis was proposed on the basis of scanty analytical evidence obtained largely in the laboratory of P. A. Levene.

It is invidious to single out Dr. Levene's work as an example of the early fumblings in the determination of nucleic acid structure and function. With the techniques available to him he could hardly have done much more. Morevover, the number of workers interested in nucleic acids was pitifully small, and thus the development of the field lacked the momentum which is generated when many workers commit their intellect and energy to a small area. And in those days biochemists shunned nucleic acids: the proteins held the center of the stage.

As Kossel expressed it in 1911:

The number of amino acids which may take part in the formation of proteins is about as large as the number of letters of the alphabet. When we realize that through the combination of letters an infinitely large number of thoughts can be expressed, we can understand how a vast number of properties of the organism may be recorded in the small space which is occupied by the protein molecule.

A great surprise was in store for those of us who, like Kossel, focused all our attention on the proteins as the repository of genetic information.

4 THE TASK OF A GENE REVEALED

ONE GENE → ONE ENZYME

WE LIVE because we have enzymes. Everything we do—walking, thinking, reading these lines—is done with some enzyme-motivated process. Life may be defined as a system of integrated enzyme reactions. A living cell functions on the harmonious cooperation of perhaps 10,000 different enzymes. In turn, an organism such as man is a constellation composed of a hundred thousand billions of such cells.

How are these enzymes made, and what happens to the whole organism if there is a failure in the making of one or more enzymes? The first, the more fundamental of the two questions, was actually answered about sixty years ago, but we were too obtuse to recognize it.

In 1902 a keenly observant English physician and biochemist, Sir Archibald E. Garrod, published a paper in Lancet which he entitled prophetically: "The Incidence of Alkaptonuria, a Study in Chemical Individuality."

The visible diagnostic characteristic of alkaptonuria is the blackening of the patient's urine when stale. (The telltale darkening of diapers or bed linen permits easy detection during infancy.) Some of the other manifestations of the disease are the pigmentation of cartilage and tendon tissue in the adult.

The agent, which turns black on prolonged contact with air, was isolated and proved to be a chemical called homogentisic acid.

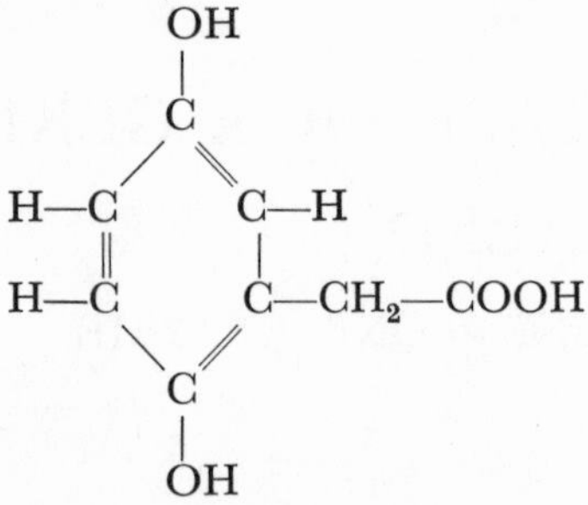

Homogentisic Acid

Homogentisic acid is present in the urine of these patients but it is absent from urine voided by normal humans. Garrod traced the genealogy of some of his patients and found that alkaptonuria is an inherited disease. It is passed down from parent to child as a recessive Mendelian character. Garrod recognized that the symptoms are the result of some aberration of metabolism and called the disease, very aptly, an "inborn error of metabolism." Homogentisic acid proved to be a normal intermediate in the metabolism of the amino acid tyrosine.

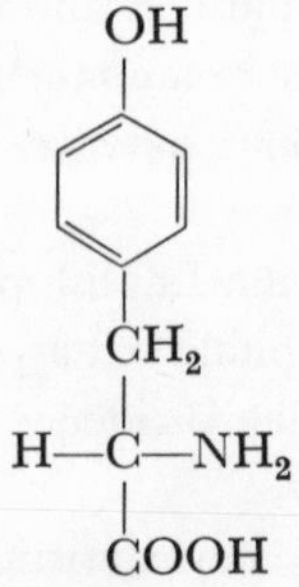

Tyrosine

For reasons which were unknown at the time Garrod made his discovery, a patient afflicted with alkaptonuria cannot metabolize homogentisic acid to carbon dioxide and water, a feat which a normal human achieves with the greatest of ease. But Garrod made a penetrating conjecture on the source of the trouble in alkaptonuria. He wrote as early as 1923: "We may further conceive that the splitting of the benzene ring of homogentisic acid in normal metabolism is the work of a special enzyme, that in congenital alkaptonuria this enzyme is wanting." In other words, since hereditary diseases stem from the gene, the failure of the gene in alkaptonuria is the failure to fashion the appropriate enzyme. This was the first statement of the seminal relationship between gene and enzyme. However, the classical geneticists were not yet ready to think in terms of molecular mechanisms. They did not visualize the gene as a specific agent but rather as some vague vitalistic entity. Therefore, the possibility that the gene is a molecular component of a cell which shapes an enzyme was not ready for acceptance.

It frequently happens in the history of science that a novel interpretation or observation leads to a host of similar findings because the attitudes of other investigators are reoriented by the perceptive pioneer. Garrod's patient research and intuitive interpretation was such a turning point. A number of different inborn errors of metabolism have been catalogued in the past sixty years since Garrod's initial observation. The ones that are recognized are those in which the crippling effects of the metabolic blunder are relatively minor, enabling the afflicted to survive after birth into infancy or even adulthood. However, there must be many others which are unrecognized because they doom the foetus or the newborn to death of undiagnosable origin.

The extensive, unexpected damage that an inherited metabolic error can produce is well illustrated by a syndrome which is variously called phenylketonuria or phenylpyruvic oligophrenia. These patients, too, excrete in their urine a substance, phenylpyruvic acid, which is largely absent from normal urine. If they survive infancy, as many do, the skull of the afflicted does not achieve normal growth—hence the term oligophrenic. The brains housed by the tiny skulls function but poorly, resulting in intelligence quotients of 20 to 50.

This particular aberration, which proved to be an inborn error, was discovered by a country physician in Norway, Dr. A. Fölling. (For some time it was known as Fölling's syndrome.) About ten years ago Dr. Fölling, although he was in his early sixties, decided to equip himself with some modern tools for the pursuit of his studies. He came to the United States on a traveling fellowship wisely provided by the Rockefeller Foundation, and, among other laboratories, visited mine to learn some techniques to determine levels of amino acids and their metabolic derivatives in the blood and spinal fluid of oligophrenic patients.

Since the history of ideas and of discoveries is a hobby of mine, this was a fine opportunity to hear the circumstances of a discovery at its source. For the sake of brevity the steps, or more appropriately the stumblings, leading to a biological discovery are usually not described in scientific communications and they are therefore frequently lost. Sometimes, especially recently, the story of the method of the discovery becomes slightly distorted by a bit of retrospective insight with which some scientists, especially some members of the younger generation, seem to be especially endowed. Needless to say, the purported method of discovery is always ascribed to the brilliant insight of the investigators. Did they but know a bit of history of science they would realize that many of the truly great discoveries were made serendipitously and that anyone making a real chance observation—which he interprets soundly—is thus in good company.

Dr. Fölling is a modest, self-effacing, dedicated man and against the background of Big Science of today his story has the charm and truth of a primitive painting against an exhibit hall full of contrived, lurid, nonobjective art. He was a country doctor in Norway, and, like many sensitive men before and after him, found the constant sight of human pain and the frustrations at the impotence of our total medical knowledge in the face of many of our diseases too great a burden to bear. One day a pretty, blond young girl who was severely mentally retarded was brought to him by the despairing mother. Dr. Fölling said he still does not know why, but just to do something he decided to add a chemical reagent, ferric chloride, to a sample of her urine. It turned green because of the presence of what later was proved to be phenyl pyruvic acid. And as Dr. Fölling put it, "God was good to me. I did not have to practice medicine since then."

Dr. Fölling followed up his observation with drive and ability and he was thus able to draw attention to an inborn error of metabolism—for such it turned out to be—with a crippling effect on the subject's mental capacity. The syndrome is not too rare. It claims one out of 25,000 infants.

Subsequent research showed that the metabolic blunder is a block in the patient's ability to convert one amino acid, phenylalanine, to another, tyrosine.

Phenylalanine   *Tyrosine*

The biological reaction which goes with ease in a normal human goes on at only one tenth the normal rate in those afflicted with phenylketonuria.

The excretion of phenylpyruvic acid results from an alteration effected by the kidney in the accumulated phenylalanine, i.e., the replacement of the NH_2 group with an oxygen.

$$
\begin{array}{c}
H \\
| \\
C \\
\diagup\diagdown \\
H{-}C \qquad C{-}H \\
| \qquad\qquad | \\
H{-}C \qquad C{-}H \\
\diagdown\diagup \\
C \\
| \\
CH_2 \\
| \\
C{=}O \\
| \\
COOH
\end{array}
$$

Phenylpyruvic acid

Since tyrosine is the source of the body's pigments, these people are usually blond because of the lack of adequate source material. From the clinical point of view phenylketonuria turned out to be a milestone because it is the first mentally crippling disease whose destructive effects can be obviated, provided the diagnosis is made early.[1] It was found that if the afflicted infants are placed on a diet that is very low in phenylalanine they escape the brain damage.

To be sure such a diet is very expensive—no protein can be used, and thus all of the other amino acids must be individually administered. However, it is one step forward in the long road that may lead to the conquest of mentally crippling diseases.

Dr. Fölling was honored in 1963 by the Kennedy Foundation

[1] In New York State it is now legally required to test a newborn infant's urine for the presence of phenylpyruvic acid.

at an impressive ceremony at the White House for his initial discovery which led to this triumph, however limited, over the scourge of the most devastating cripplings, the crippling of the master organ, the brain.

The inborn errors of metabolism should have served as a clear clue to the mechanism of gene action. They are hereditary diseases, therefore they are transmitted by the presence, or, more likely, the absence of a gene. If a phenylketonuric cannot convert phenylalanine to tyrosine it must mean that the appropriate enzymes are lacking, since all biochemical processes are achieved by those wondrous catalysts of the cell. But this is obvious in hindsight only. And hindsight, contrary to real sight, sharpens with the passing years.

Not every enzyme is of pivotal importance for the life of an organism. For example, we humans have lost the necessary enzymes for the manufacture of ten out of twenty amino acids. During our evolutionary ascent we also lost the enzymatic capabilities for the shaping of vitamins as well. The loss of those genes and enzymes was not deleterious, for having become predators, we could easily compensate for them by eating other organisms, plants, and animals that retained these synthetic capabilities.

However, with advanced cultural development, as we humans became less proficient predators and failed to compensate for it by increasing the sources of food with adequate farming and animal husbandry, those lacunae in our enzymatic armory became increasingly felt. That primitive peoples suffered from nutritional deficiencies is testified to by the stigmata of the diseases in their remnant bones and by some of the stories in their religious lore. But only within the past sixty years have we correlated certain diseases with the deficiency of vitamins in the diet. Still more recently are we aware of the organic symptoms of malnutrition induced by amino acid deficiency. Kwashiorkor, a disease caused by deprivation of adequate proteins, is widespread among large segments of the human population.[2] Mankind must increase enormously its protein-producing capacity or it must control its own proliferation to keep the essential balance imposed by the loss of the appropriate genes during our evolutionary ascent.

Obviously all gene losses and, consequently, their enzymes are not equally harmful. The loss of enzymes for the making of vitamins has not been an insurmountable blow. (Especially is this true today when we can manufacture vitamins by the hundreds of tons.) Nor is the loss of the ability to form body pigments incapacitating. Albinism, which is prevalent among all mammals, is but a minor handicap among humans. (They sunburn too easily.) Among whales albinism is probably no handicap at all. (If Melville is to be believed, Moby Dick was huge in size, prodigious in endurance, and fierce in temper. For despite what the literary psychoanalysts try to make him out to be, Moby Dick was just an albino whale.) The loss of the enzyme which converts phenylalanine to tyrosine in phenylketonurics is a more severe handicap. The accumulating products do irreparable damage to the brain. The more peripheral the function of an enzyme, the less damaging is its loss to an organism. The disappearance of a pivotal enzyme dooms the creature in which the mutation occurs. Such mutations are lethal and therefore not perpetuated. Obviously, no organism could survive a moment if it lost the ability to make adenosine triphosphate—the ubiquitous source of energy for all biochemical processes.

The information from inborn errors of metabolism failed to reveal to us the relationship between genes and enzymes, but fortunately biological truth has a way of struggling to the surface

---

[2] The word *kwashionkor* means "redheaded boy" in some African dialect. A red pigment deposits in the hair of those afflicted with the ailment. The aberrant metabolism which produces the pigment—and provides the exotic name for the syndrome—is not an inborn error of metabolism for it is produced by a deficiency, not in the subject's genetic makeup, but rather, in his environment.

provided there are gifted men to help it. One such man is Dr. H. J. Muller, a student of Morgan's.

Dr. Muller is one of those few scientists who can state a scientific problem not only in the arcane code of scientese, but in a style that is clear to everyone, so I shall quote his statement of the task he undertook.

Most modern geneticists will agree that gene mutations form the chief basis of organic evolution, and therefore of most of the complexities of living things. Unfortunately for the geneticists, however, the study of these mutations, and, through them, of the genes themselves, has heretofore been very seriously hampered by the extreme infrequency of their occurrence under ordinary conditions, and by the general unsuccessfulness of attempts to modify decidedly, and in a sure and detectable way, this sluggish "natural" mutation rate. Modification of the innate nature of organisms, for more directly utilitarian purposes, has of course been subject to these same restrictions, and the practical breeder has hence been compelled to remain content with the mere making of recombinations of the material already at hand; providentially supplemented, on rare and isolated occasions, by an unexpected mutational windfall. To these circumstances are due the widespread desire on the part of biologists to gain some measure of control over the hereditary changes within the genes.

In order to increase the very low rates of natural mutations—which are on the order of a few per million—Muller's predecessors exposed organisms to alcohol, lead, and other toxic substances. The results were either negative or of doubtful significance. Muller decided on stronger measures. In 1927 he exposed the fruit fly to the intensely penetrating energy of X rays.

He was successful beyond his dreams. He found that the spontaneous mutations which had been observed previously among populations of fruit fly—identified, for example, as "miniature wing" or "forked bristles"—appeared with much greater frequency among the offspring of populations exposed to X rays. But more than this, he found numerous mutations theretofore unobserved, such as "splotched winged" and "sex combless" flies.

The gene was penetrated at last. The intense energy of the X ray on impact with a gene target dislocates its structure and thus produces permanently monstrous offspring. For Muller found that the visibly altered structures in the progeny of the flies exposed to X rays continued in evidence in subsequent generations. The X rays brought about a permanent genetic alteration, in brief, a mutation.

Important though Muller's discovery was—he received the Nobel Prize for it—it still told us nothing about the mechanism of gene action. A concatenation of several factors might yield a fly with "miniature wings." But Muller's discovery eventually enabled us to study at the molecular level how the invisible gene controls the shape, the color, the very essence of a living thing. A brilliantly successful partnership between Dr. G. W. Beadle, a geneticist, and Dr. E. L. Tatum, a biochemist, was formed in 1938 to explore the problem. These fruitful partnerships and the discoveries which bloom from them are not the product of chance. Discoveries are not only the achievement of the particular scientist, they are the product of their time. As Dr. Tatum put it in his Nobel address, a discovery depends on "knowledge and concepts provided by investigators, past and present all over the world; on the free interchange of ideas within the international scientific community, on the hybrid vigor resulting from cross-fertilization between disciplines; and last but not least on chance, geographical proximity and opportunity." I would add one further ingredient which, as in the case of good generals, plays a role in the achievement of good scientists: luck.

In this team Dr. Beadle was the biologist, but a biologist who was not content with mere observation of biological phenomena. He yearned to intercede and try to understand biology at the level of molecular mechanisms. In particular, he was eager to probe the mode of action of a gene. While at the California Institute of Technology he joined forces in 1933 with a visiting French biologist Dr. Boris Ephrussi, who was also impatient with the essentially descriptive approach of classical biologists to the basic problems of genetics. The two of them decided to explore a genetically controlled attribute of the fruit fly, its eye color. Normally the eye color of the fruit fly is brown, and this was known as the wild-type eye color, but by the mid-1930s some twenty-six other eye colors were known. These colors—vermilion, cinnabar, claret, etc.—are transmitted to offspring and must therefore be under the control of genes. Beadle and Ephrussi decided to transplant eyes from fly to fly in the larval stage before the colors had developed. With infinite patience they perfected their techniques until they could consistently create three-eyed monsters. Inspection of the developed trioptic creatures revealed an exciting discovery. The eye from a wild-type larva which was destined to be brown—had it been permitted to remain attached where it belonged—did not develop the brown pigment in its mutant foster larva. On the other hand, in the reverse case, an eye which normally would be vermilion developed brown pigment in its wild-type foster home. Beadle and Ephrussi interpreted their finding with profound insight. They concluded that in the wild-type larva there must be some substance which can diffuse into the transplanted eye and convert it to the same brown color as the other two eyes. On the other hand, this pigment-forming substance must be absent from the vermilion-eyed host. That there must be some variants of this pigment-forming substance was shown when they found the following subtleties: A vermilion eye in a cinnabar host became brown, but a cinnabar eye in a vermilion host remained cinnabar. The investigators concluded that the inability of the flies with the exotic eye colors to develop the usual brown color must be due to a failure of their genes to produce the appropriate precursor for brown pigment formation. The subtle differences in the cinnabar and vermilion hosts were ascribed to sequential changes of the pigment precursor in the different mutants. These conclusions are still valid today. Dr. Beadle came to the conclusion that genes act by regulating chemical events. He therefore decided that he needed a chemist to extend his findings. He chose well.

Dr. E. L. Tatum had sound training at the University of Wisconsin in chemistry and biochemistry and in the role of vitamins in the nutrition of microorganisms. He also spent a year as a post-doctoral journeyman at the University of Utrecht in the laboratory of F. Kögl, who had just discovered biotin, a new member of the vitamin B group.

Dr. Tatum joined Dr. Beadle, who had by then moved to Stanford University. (American scientists often move around from post to post like so many migrant workers, as universities keep bidding for the better ones among them.) The first task to which they addressed themselves was a study of the nature of the eye-pigment precursor. Dr. Ephrussi had reported from Paris that if the amino acid tryptophan is incorporated into the diet of the fruit flies with mutant eye color there is a trace of brown pigmentation of the eye. Tatum took off from there and grew mutant fruit flies in the presence and absence of tryptophan under sterile conditions, and here is where Lady Luck waved her wand on Dr. Tatum for the first time. The tip of her wand was dipped in a culture of bacteria. The abode of one of Tatum's vermilion-eyed mutant cultures containing tryptophan became infected and the microorganisms altered the tryptophan in such a way as to render the amino acid an excellent source of brown pigment. Tatum isolated the substance produced by the bacteria and identified it as kynurenin, whose structure, shown on the next page, reveals that the smaller ring in the parent amino acid is broken.

Tryptophan

Kynurenin

This is as far as Beadle and Tatum went, but Adolf Butenandt, an excellent German organic chemist who is interested in the structure of natural products, carried the problem further and showed still another precursor of eye coloration—3-hydroxy-kynurenin.

3-Hydroxykynurenin

The eye-color sequence was completed by Butenandt, and briefly it might be written thus:

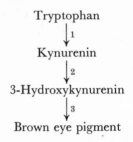

$$\text{Tryptophan}$$
$$\downarrow 1$$
$$\text{Kynurenin}$$
$$\downarrow 2$$
$$\text{3-Hydroxykynurenin}$$
$$\downarrow 3$$
$$\text{Brown eye pigment}$$

The flies which can develop brown eyes normally can carry out all three of the above conversions. A vermilion-eyed mutant cannot carry out reaction 1. Therefore, if it gets kynurenin instead of tryptophan it can proceed to the complete pigment formation. A fly with the cinnabar eye coloration can perform reaction 1 but not 2.

Work with the fruit-fly eye color was frustratingly tedious for Beadle and Tatum. (It took Butenandt a decade to identify 3-hydroxykynurenin as a stage in eye pigment formation.)

A resilient flexibility in approach to a problem often distinguishes the great scientist from the merely competent. Beadle and Tatum decided to explore the mechanism of gene action in microorganisms. They were influenced in their decision by some penetrating deductions by an outstanding French biologist, Dr. André Lwoff, of the Pasteur Institute. Lwoff had studied the nutritional requirements of certain parasitic microorganisms, both protozoa and bacteria. He observed that many of these parasites often lose the ability to synthesize some of their essential nutrients, which they must siphon off from the hosts they infest. Lwoff clearly associated the loss of synthetic capacity with loss of the appropriate genes, for he spoke of the loss of some synthetic abilities as part of an evolutionary trend. And, of course, evolution operates through the selection of mutants.

The name of the organism Beadle and Tatum chose,

Neurospora crassa, or bread mold, was recondite in 1940, but today it is a household word in biology. In making the choice the two investigators reached into their own scientific backgrounds and decided on this organism through an extraordinary set of coincidences. In the first place they wanted an organism in which the basic genetic work had already been done. Dr. Beadle recalled that while still a graduate student at Cornell he heard B. O. Dodge of the New York Botanical Garden give a seminar on genetic studies with Neurospora. Dr. Dodge kept on insisting even to T. H. Morgan that Neurospora is better than the fruit fly for genetic work. (Dr. Dodge, who lived to be an octogenarian, was around to see his prophecy fulfilled.) Tatum in turn brought a good deal of personal knowledge to the task. He had worked on the nutritional requirements of microorganisms, but more to the point he had shared a laboratory in Utrecht with Dr. Nils Fries, who had perfected a completely synthetic medium on which molds such as Neurospora could be grown. As we shall soon see this was an enormous asset.

The bread mold, Neurospora, makes very few demands on the world. It thrives on water, sugar, inorganic salts, and just a trace of one of the B vitamins, biotin. The reader may recall that Kögl, in whose laboratory Tatum spent a year, had already isolated biotin. It was available in pure form and thus there was no fear that unknown vitamins or other nutrients might sneak into the diet prepared for Neurospora.

The strategy of the experiment was simple. Out of the few starting materials in its diet Neurospora fashions everything it needs for the growth of its cells. It makes its own amino acids, all twenty of them; it synthesizes all of its vitamins, except, of course, biotin; almost every cellular component of Neurospora is homemade.

The organisms were exposed to X or ultraviolet irradiation and were planted into culture media into which all of the known vitamins and amino acids had been incorporated, and the survivors of the irradiation were permitted to grow in this nutritional paradise. Spores of the organisms which thrived on the enriched medium were transferred to the original minimal medium on which their unirradiated ancestors—these are called the wild type—could grow with ease. To the delight of the investigators, some of the transplants did not thrive in the minimal medium: they had apparently lost, as a result of the irradiation, the ability to fashion some of their nutrients. Scores of new media were prepared in which the bare essentials were supplemented with a vitamin or an amino acid and the spores which were doing poorly in the minimal media were transferred to a partially supplemented environment. Several were found which thrived if only one amino acid or one vitamin was available to them. And the requirement for the extra nutrient persisted after many generations. The experiment, so brilliantly conceived and faultlessly executed, was thus crowned with well-deserved success. Mutational changes were induced which were not expressed by gross alterations in the appearance of organs, as in the case of complex organisms, but by some invisible lesion within the irradiated cell whereby the capacity to synthesize a specific nutrient was lost. The irradiated Neurospora were crippled just as the irradiated fruit fly's offspring were, but in this case we knew exactly what was wrong. A mutation occurred that destroyed a specific capacity for a specific synthesis. But all chemical syntheses within the cell are operated by enzymes, therefore Neurospora revealed that a mutational change is a loss of some enzyme. The function of the gene now became clear: A gene must somehow give rise to a specific enzyme. Of course this is exactly what Garrod had stated some seventeen years earlier.

As a result of the work of Muller and the brilliant choice of experimental material by Beadle and Tatum, we need no longer wait for chance mutations in animals; we can induce them in microorganisms and study the particular inherited lesion, the absence of a specific enzyme.

The second sweeping conclusion which was drawn from this work is that genes, wondrous though their collective capability may be in creating a whole organism, be it mold, mice, or men, are individually highly circumscribed in their capacity. A particular gene can make only one specific enzyme.

Identical experiments were performed on bacteria, among which a great many nutritional requirements were known to occur naturally. The experiments were again successful beyond expectation. Bacterial mutants with nutritional requirements ranging from vitamins to nucleic acid precursor bases were soon produced by the scores. For the work with the bacteria Dr. Tatum chose a typical strain of the colon bacillus from the collection in the Bacteriology Department at Stanford. He was handed a strain called *E. coli* K_{12}—and as we shall soon see he was still a favorite of Lady Luck.

As news of these exciting developments spread, investigators young and old trooped to the new frontier and soon new mutants by the scores were being described. A so-called biochemical mutant was found for almost every amino acid, every vitamin, and every nucleic acid precursor base. An unexpected bounty from these mutants is the development of easy assays for these nutrients. A bioassay for a vitamin, say vitamin B_1, using birds as the experimental animals, is tedious, time consuming, and smelly. It takes weeks to deplete the vitamin reservoirs in the animals, the preparation of the test diet is exacting and expensive, and some animals will be ornery and refuse to come down with the expected vitamin deficiency. If a microorganism is available which is a biochemical mutant for vitamin B_1, in other words it requires B_1 in its diet because it lost some gene for the synthesis of this vitamin, then the assay becomes simple. The number of bacteria which can grow in a given volume is proportional to the amount of the nutrient present in limiting amounts. Thus it can be easily established, for example, that 10^8 bacteria can grow on X micrograms of vitamin B_1. Therefore if a test sample of vitamin B_1 supports the growth of 2×10^8

bacteria, then the sample must contain 2X micrograms.   I will
not impose on the patience of the reader by a description of the
technical details of such an exercise.

But far more important insights into the living process became
ours as work with biochemical mutants expanded.   It soon
became apparent that there can be several different mutants that
have the same nutritional requirement.   For example, several
mutants of a microorganism may be isolated, each of which
requires the amino acid tryptophan.

Tryptophan

But in one of them a compound called indole can replace the
completed amino acid tryptophan.

Indole

In another mutant an even simpler compound, anthranilic acid,
will serve in lieu of the tryptophan.

$$
\begin{array}{c}
H \\
|\\
C \\
HC \quad C \quad COOH \\
HC \quad C \\
C \quad NH_2 \\
|\\
H
\end{array}
$$

Anthranilic Acid

In a third mutant a still more primitive compound, shikimic acid, is adequate.

$$
\begin{array}{c}
H \quad H \\
HO \quad C \quad COOH \\
HC \quad C \\
HC \quad H \quad C \\
HO \quad C \quad H \\
|\\
OH
\end{array}
$$

Shikimic Acid

These various mutants revealed that a substance like tryptophan is not whipped together by a single knowledgeable enzyme. For potent though enzymes are, their versatility is invariably highly circumscribed. These mutants charted the plodding path of synthesis of tryptophan by a sequence of enzymes, each of which adds its bit to the total task.

There may be as many as a dozen sequential steps in the assembly line that makes a product, such as an amino acid. Each of those steps is enzyme motivated and therefore a dozen genes must exist which carry the information for the shaping of those enzymes. If but one gene is lacking or is imperfect, the assembly of the amino acid cannot be completed. Should the missing enzyme be the one which completes the very last step of the synthesis, then the organism can survive only if the completed amino acid is supplied to it from an outside source. However, if the missing enzyme fabricates a part of the amino acid early on the assembly line, then any substance which comes after that step can substitute for the amino acid. Since all the other enzymes are present they will work on the material which is proffered to them just as if it had been made by the missing enzyme, and thus complete the production of the amino acid.

Actually, we frequently find that there is only one gene and consequently one enzyme lacking in the sequence of enzymes that yields a needed product. This can result in a curious symbiosis between two of the crippled microorganisms. In any biochemical mutant there will be an accumulation of the metabolic product which is produced just prior to the blocked reaction. This is what is to be expected of any manufacturing process which is based on a sequential, assembly-line-like procedure. The reader may recall the machine production line in Charlie Chaplin's inspired movie *Modern Times*. Just as there was a pileup on the assembly line produced by the inept little man with the mustache, so there is a pileup within a cell if an enzyme is lacking for the next operation. When the exquisitely delicate balance of products within a cell is upset, the cell resorts to its first line of defense against any toxic product: it tries to excrete it. This is true in man, as it is in microorganisms. The syndromes alkaptonuria and phenylketonuria, described by Garrod and Fölling, were detected by the appearance of the piled-up metabolic products in the urine of the afflicted. It should be emphasized that there is nothing wrong with the kidneys of such subjects; indeed their kidneys are striving heroically to eliminate the excess and therefore toxic metabolite which has piled up in the bloodstream. For it is an anomaly stemming from the infinite intricacy of the machinery of a cell that metabolic products which normally are seized upon and fashioned into needed compounds become nefarious toxic irritants when they reach excessive levels. The puritan-inspired adage about too much of a good thing certainly applies to a living cell's metabolic economy.

Microorganisms, too, excrete partially completed products which may accumulate if there is a block caused by a missing enzyme in their metabolic assembly line. And this brings about the possibility of symbiosis between two or more metabolically crippled microorganisms.

It is technically very simple to grow biochemical mutants of microorganisms in large batches, and therefore the isolation of their accumulated products is relatively easy. The cataloguing of the intermediate stages in metabolism which accumulate in enzyme-crippled mutants, together with information from the use of isotopic tracers (see Chapter 7), revealed an astonishing chemical complexity within a living cell. To make the twenty amino acids it needs, a Neurospora or a bacterial cell must have about a hundred different enzymes. Still more enzymes are needed to fashion the vitamins, the nucleic acid precursor bases, and other needed components of the cell. About forty different enzymes are needed just to release the energy locked in a sugar molecule and convert it into a form which the cell can use to drive the many enzyme reactions which consume energy for their completion. Hundreds of other enzymes are needed to weave together the many small building units into the exquisitely structured giant molecules, the proteins, the nucleic acids and the complex sugars. Still other enzymes are needed to form the membranes and organelles that make up the wondrous world of a cell. All of these enzymes, which total perhaps 10,000, must act in harmonious concert to perpetuate that most precious product of molecular and organic evolution, a living cell.

Whether an organism can survive the loss of a gene depends on how pivotal is the gene's product, the enzyme, in the cell's total economy. It is interesting in this connection that, though many mammals have lost much of their amino acid and all of their vitamin-synthesizing capacity, all have retained through their long evolutionary ascent the ability to make all of the building units of their nucleic acids. In Nature only a few parasitic protozoa are known which lost some nucleic acid precursor-synthesizing ability. It would appear that these building units are too important to have their steady supply left to the uncertainties of the diet. Loss of these synthesizing capacities must have been too great a handicap, and such biochemical mutants must have been left as derelicts on the roadside in the surging path of evolution.

The work of Beadle and Tatum revealed the mode of action of genes and thus the mechanism of mutation; it contributed enormously to our armory of knowledge of intermediate pathways of metabolism and made still another contribution: it revitalized the science of genetics. In this area Dr. Tatum had yeoman help from a student, Dr. Joshua Lederberg. He was a first-year medical student at Columbia who chafed at merely sitting at the delivery end of a large funnel through which his professors were pouring the copious amounts of descriptive information on which the structure of medical knowledge is built. A man of enormous energy, Lederberg supplemented his medical studies with a variety of scientific enterprises. He studied basic chemistry, he dabbled in biochemical research and in Neurospora genetics. With the encouragement of his mentor in the latter area, he quit medical school and went to work for a doctorate in science with Tatum, who had by then migrated to Yale. There he was handed undoubtedly the best Ph.D. problem in the history of biology on a silver, no, not platter—on a silver Petri plate.

The basic question Tatum was asking was whether there is sexual union and consequently genetic fusion in bacteria. Bacteriologists had observed under the microscope rare pairs of bacteria in suggestively close proximity. But in the past these were brushed aside as random artifacts stemming from the immense population density of bacterial cultures. To probe the sex life of bacteria Tatum's strategy was impeccable. He visualized mixing together two different biochemical mutants of bacteria, each with a different genetic and therefore enzymatic deficiency. If there is genetic fusion, then from the pooled genetic resources of the two genetic cripples the original, wild-type organism which had the ability to make both nutrients should be reconstituted. For example, let us assume that one irradiation-produced mutant, A, cannot make the amino acid methionine; another mutant, B, lost the gene for making the amino acid leucine as a result of irradiation. When the two bacteria are mixed the ability to make leucine might pass from A into B (which had retained its methionine-making ability all along) and therefore the offspring of such a union would be able to make both amino acids.

Tatum gave Lederberg two mutants of *E. coli* K_{12} which he had isolated at Stanford. The one that required methionine had been arbitrarily designated as 58–161. The one that required leucine is called W–1177. The technique to be followed could not have been more simple. If either of these mutants is spread on a sterile Petri plate of agar into which have been incorporated glucose, inorganic salts, and the requisite amino acid, the bacteria grow happily. But if the essential amino acid is omitted from the diet, the organism will not grow. If as many as a hundred million bacterial cells (10^8) are spread on such an unsupplemented plate, perhaps ten bacteria will be found which do not need the missing amino acid and these will multiply in isolated areas, giving rise in thirty hours to visible clumps of bacteria. (If a bacterium divides every hour, in thirty hours there will be about a billion of them and in a single clump, called a clone, such a number is easily visible with the unaided eye.) But if 58–161 and W–1177 are intimately mixed and are plated on the same minimal agar plate, instead of ten visible clones there will be hundreds, resulting from the pooled genetic know-how of the two different mutants.

The progeny from these clones will continue to grow indefinitely, independently of any added amino acids.

As our knowledge of sexual reproduction in bacteria emerged, we looked with awe at the luck that continued to hover over Tatum's experiments. There are few strains of coliform bacteria that undergo sexual recombination. *E. coli* K_{12}, which was handed routinely to Dr. Tatum out of Stanford's collection of such bacteria, is one of those which participates in sexual congress. (Figure 4.1 shows two such bacteria—under an electron microscope—exchanging their genetic material through a mating tubule, which forms between them. The small hexagonal shapes are dead bacteriophages which were permitted to come in contact with the female to identify her.)

Moreover, it was shown later by Dr. Lederberg that there is a definite sexual polarity among bacteria, and successful exchange of genetic material can occur only between an organism with a male and one with a female orientation. The nutritional mutants produced by irradiation are almost invariably female, and, therefore, with most pairs of amino acid-requiring mutants the genetic recombination does not go. The methionine-requiring mutant (58–161) happened to be a rare male, so the very first attempt to mate the bacteria—and the experiment—was fertile.

My emphasis of the role of chance in the success of these experiments is not intended to detract from the brilliance of the achievement. Tatum's concept was nothing less than revolutionary and the experiments were faultlessly designed. Indeed, I am the last to envy Dr. Tatum for the good fairy who smoothes his experimental path. She spared a stroke or two of her brush for me also. With the aid of 58–161 my students and I were able to show the existence of a whole new group of enzymes which are present in every living cell whose function is the alteration of the structure of nucleic acids to confer on them an individuality characteristic of that species of organism. Good old 58–161 is

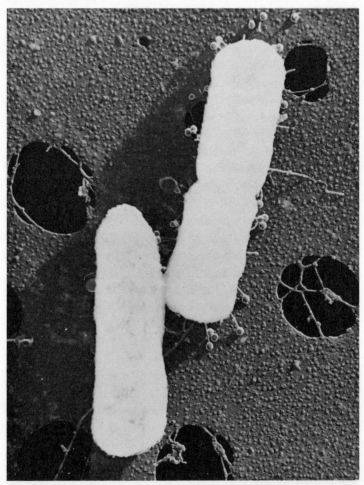

Fig. 4.1. Two bacteria during the act of conjugation. The female was marked by placing her in a suspension of killed bacteriophage. The latter become attached to their victim but are otherwise innocuous.

the only mutant known which could make such a revelation possible.

Once Dr. Lederberg was on his own he pushed forward with driving energy and inspired originality. He laid down the groundwork for a new science: bacterial genetics. Bacteria are extraordinarily well suited for genetic studies. The vast numbers which can be used in an experiment bring statistical reliability to a finding. (There can be more bacteria in a teaspoonful of liquid than there are humans on this globe.) The speed of their growth provides answers to questions with gratifying swiftness, The coliform bacteria divide in anywhere from twenty to sixty minutes, depending on the kind of diet they are offered, and thus a single bacterium in which a mutation has occurred can have a billion descendants in about thirty hours. Questions which would require months to answer with the fruit fly as the genetic tool can be resolved in days with bacteria. For example, we can tell in a couple of days whether a certain chemical compound can produce mutations. (After it was shown that X rays and ultraviolet irradiation cause mutations it was found that many chemicals are also mutagenic.) We simply determine the frequency of a certain biochemical mutant in a population of bacteria before and after exposure to the chemical.

Bacterial genetics can answer questions that are beyond the reach of any other discipline. For example, there was the puzzling problem of the gradual resistance of bacteria to drugs. Soon after the antibiotics were introduced into clinical practice it became apparent that bacteria which infest a patient become increasingly resistant to a given antibacterial agent. After repeated use of the same antibiotic—especially if the doses are inadequate in level and duration—it becomes useless against a given infection. The problem posed is simply this: Is resistance to a drug developed by the drug itself or are there some bacteria which are resistant before ever encountering the drug? Lederberg answered this question with an experiment which is so simple both in conception and in design that everyone of us berated ourselves for not thinking of it first. But a retrospective obviousness is the hallmark of every sound hypothesis or experiment. The outstanding scientists are those who can see the obvious. Or, there are those who can see what everyone else has seen but who have the insight and courage to give a new interpretation to it. As the French philosopher-scientist Poincaré stated it, we are seeking in Nature *la simplicité caché*.

Lederberg's experiment is performed as follows. Let us take an agar plate on which there are about 200 clones of bacteria. We wish to determine whether among these there are any which are resistant to, say, streptomycin *before* they come in contact with this antibiotic. A piece of velvet is trimmed to the size of the agar plate and is sterilized in a high-pressure autoclave. The sterile velvet is gently pressed on the surface of the bacterial clones and it is then pressed onto two new, sterile, agar plates, A and B. In plate B, in addition to the usual bacterial nutrients, streptomycin in lethal dose is also incorporated. The spicules in the cloth will carry with them small numbers of bacteria from each clone, and some of these will be deposited on the sterile agar of the two plates. The geography of the bacterial seeds on the two plates will be identical because the velvet is carefully oriented in the same direction on each plate. The technique is aptly called replicate plating. The two plates are placed in a warm room where the bacteria sown on plate A will multiply, and in twenty-four hours this plate will be a duplicate image of the original plate; there will be 200 clones with the same geographical distribution as on the original master plate. Plate B will present an entirely different sight. Instead of 200 clones there might be only two, all the other seed bacteria having been killed by the streptomycin. The two resistant clones may be at, let us say, twelve o'clock and three o'clock on the round plate. The twin counterparts of these two can be easily identified on plate A and we can now determine whether or not the two clones on plate A which had not been in contact with streptomycin are sensitive to the drug. These clones can be lifted off with a sterile wire loop and transferred to new plates C and D which contain the lethal concentration of streptomycin. The transferred clones are spread all over the plates with the wire loop, in effect layering the plates with millions of seed bacteria. After twenty-four hours of incubation in the warm room these plates will have the answer to our important biological query. This is the moment of truth in the life of the scientist. The challenge is infinitely more complex than that faced by the bullfighter, for whose supreme moment the above cliché was coined. But the matador faces merely brute force and his truth is at once trivial and ephemeral. He proves that he has the raw courage and the agility of feet to put his life on a scale whose other pan is filled with a pile of money. The scientist's motives are usually equally self-centered. Curiosity about the world of which he is a part is probably the most universal motivation, proof of his capabilities vis-à-vis his colleagues is a pervasive one, and, especially recently, careerism is creeping in as an increasingly prevalent motivation. The truth the scientist seeks, while not necessarily eternal, should be durable enough to withstand the onslaughts of his critical peers. We try to enlighten our minds by evidence that can survive scrutiny and critical repetition. In the biological sciences we seek a tiny glimpse of a miniscule segment of the infinite maze that is a living cell. Our moment of truth is our ultimate reward. Scientific work is often repetitive, tedious drudgery. Accidents which demolish days of work are frequent, many of our ideas are barren, findings often turn out to be spurious. But occasionally an idea occurs to us which has a compelling ring of truth, and the tedium and discouragement are banished by the anticipation of the answer. At least one scientist I know would run up five flights of stairs to look at plates like C and D, which we left to incubate at the start of this digression on scientific motivation.

Plates C and D would be covered by uncountable numbers of

bacteria, all of which, therefore, must be streptomycin resistant, as, of course, were their ancestral clones which had never been in contact with the drug. The prototype of this experiment performed by Dr. Lederberg revealed the mechanism of the proliferation of drug-resistant organisms. In every population of pathogenic organisms there are a few that are impervious to a given drug even before it is applied. When most of the pathogens are eliminated not by the patient's defense mechanisms but by the drug, the resistant ones survive, and, unhampered by competition with the eliminated ones, proliferate. The crisis is temporarily eliminated, but the patient may become a walking incubator of drug-resistant organisms. However, the situation is not as hopeless as this sounds. Chemists and microbiologists in pharmaceutical houses manage to keep coming up with new antibiotics to thwart the resistant little beasts.

The replicate plating technique provided enormous help in the search not only for drug-resistant but for biochemical mutants as well. Such a search used to be tedious and time consuming, but now it is swift and systematic. The search for biochemical mutants is a nice example of how quickly scientific methodology burgeons. Suppose we need a mutant microorganism which cannot make the nucleic acid precursor-base thymine. Here is how we would go about it. A suspension of the original, or wild-type organisms in minimal medium would be exposed to a dose of ultraviolet irradiation which would kill 99 percent of them. Penicillin would then be added and the survivors would be incubated. This step is rooted in the mechanism of killing of bacteria by penicillin. Only growing organisms are killed by this antibiotic. None but the original wild-type organisms can grow on the minimal medium: the nutrients for the newly made mutants are lacking. The wild-type bacteria are thus eliminated and the concentration of the mutants is enhanced. The survivors would be scrubbed free of the penicillin and permitted to multiply in a medium that contains thymine. About

200 of these bacteria would be transferred to an agar plate which contains the minimal medium plus thymine and would be permitted to grow up to clones. Many of these clones will still be of the wild type which had escaped the penicillin. Perhaps one or two may be real thymine requirers: replicate plating will reveal which ones. An impression of this master plate would be transferred to two different agar plates, one with thymine and one without. Any clone which grows on the thymine-supplemented plate but *not* on the unsupplemented plate stems from a thymine-requiring biochemical mutant.

The ease of selection enabled workers in the field to collect organisms with multiple deficiencies. Thus there may be mutants which require a couple of amino acids, a vitamin, and have in addition certain known drug resistances. The existence of organisms with—as they are called—multiple genetic markers enabled two Frenchmen to answer a basic question about the organization of genes within a chromosome.

How are genes arranged? Are they clustered together as grapes in a bunch or are they arranged sequentially as messages on a ribbon issuing from a teletype machine? Are they placed in a definite order or are they scattered haphazardly? And what about those genes whose tasks are related, as in the case of genes concerned with the sequential steps in the synthesis of the same amino acid. Would these be close together or scattered?

Drs. Francois Jacob and Elie Wollman of the Pasteur Institute in Paris saw a possibility of answering these questions with a simple biological experiment. As I wrote earlier, when populations of male and female bacteria are brought together, mating tubules are promptly formed. Jacob and Wollman decided to explore the strength of the mating tubules. They found that they are easily sheared if the nuptial cultures are placed in a Waring blendor. (This invention of the orchestra leader Mr. Fred Waring is as much a part of biological laboratories as of ice-cream parlors.) The fragility of the mating tubules in the shearing forces of the swirling waters opened a way to answer all of the above questions. For Jacob and Wollman were able to time the sequence of passage of various genes from the male to the female. In its simplest outline the experiment they performed is as follows. Let us assume that there is a female organism in which four different genes and consequently four different enzymes—E_1, E_2, E_3, and E_4—are lacking. The quadruply crippled female culture is mixed with an excess of wild-type males in which these four genes are intact. At intervals of ten minutes small volumes of the mating cultures are poured into sterile Waring blendors and the nuptial activities are barbarously interrupted. The wild-type males are killed by exposure to penicillin and the surviving biochemical mutants are analyzed by the appropriate replicate plating techniques. The outcome may look somewhat like this. After ten minutes of mating, most of the mutants still have quadruple requirements. Therefore, no gene had passed over during this time. After twenty minutes of mating, the mutants have three nutritional requirements, E_2, E_3, and E_4. Therefore, gene 1 must have passed over between ten and twenty minutes. The rest of the experiment would establish that gene 2 passed between twenty and thirty minutes and so on. The timing for any one marker is always found to be the same in the same strain of microorganisms. The genes, therefore, must be always in the same relative position to each other on a given chromosome.

These interrupted mating experiments revealed a great deal about genes. They are linearly aligned in well-established sequence, like beads on a rosary. The several genes which are involved in the sequential synthesis of a given product often are located very close to each other. Separate genes exist for the creation of such attributes as resistance to a drug or susceptibility to a certain virus which may infect that strain of microorganism.

In fairness to the hosts of classical geneticists who have mated and scrutinized myriads of fruit flies, it should be pointed out that many of the conclusions drawn from bacterial genetics were known from their tiring labors. But genetic studies of bacteria offer advantages beyond the reach of any other tool. Since we have in hand homogeneous populations of undifferentiated cells, the extraction of enzymes is infinitely easier than from complex organisms. Genetic work with bacteria coupled with intensive studies of the mechanics of enzyme formation revealed the existence of intricate regulatory mechanisms. There are substances in cells so far unidentified chemically which can suppress the formation of enzymes. Moreover, it became apparent mostly from the work of Dr. Jacques Monod of the Pasteur Institute that there are genes that regulate the functioning of other genes. Thus a new frontier in biology was opened up by Drs. Beadle, Tatum, and Lederberg, who shared in 1958 the highest accolade to which a biological scientist can aspire, the Nobel Prize in Medicine and Physiology.

The measurement of variation in enzymes and other body components both normal and pathological has become a versatile tool in the working kit of anthropologists who study the migration of human populations. For example, variations in blood types can be very high among different racial groups. Among the Toba Indians of Argentina and the Sioux in South Dakota the frequency of blood type O is over 90 percent and blood type AB is 0 percent. Among Asiatic Indians the frequency of blood type O is only 30 percent and AB is 8 percent. The two groups of Indians of the Western Hemisphere, although geographically far apart, are clearly related, but they are unrelated, except by Columbus' erroneous nomenclature, to the Asiatic Indians.

The folklore of Gypsies has claimed an Indian origin. Genetic analysis of their blood types bears out the validity of this claim. The distribution of the blood types of Hungarian Gygsies differs significantly from that of Hungarians and agrees quite well with that of a group of Indian soldiers who were tested. Thus the racial origin of the Gypsies is one more evidence of the insanity of Hitler's racist policy. The Gypsies were marked for extinction as a mongrel race and yet their blood types give testimony to minimal intermingling with surrounding peoples, otherwise their ancestral blood types would have been diluted. And, of course, by definition, since they are of Indian origin, the Gypsies belong to the "Aryan race."

A widely prevalent inborn error of metabolism which lay hidden until World War II has become an interesting genetic tool for the study of migrant populations. Early in the war the Japanese overran the major sources of cinchona bark, which is the raw material for the extraction of the antimalarial drug, quinine. The intensive research campaign which was organized to develop substitutes yielded several effective synthetic antimalarial agents. However, administration of one of these, primaquine, caused severe disintegration of the red cells among a small percentage of our soldiers. The unexpected fragility was investigated and it was discovered that the red cells of individuals who suffered from primaquine sensitivity were low in an enzyme called glucose-6-phosphate dehydrogenase. This is one of the many enzymes which are involved in the metabolism of glucose: as its name implies the function of the enzyme is the removal of hydrogen atoms from a glucose molecule to which a phosphate group is attached in the -6- position. There are adequate amounts of the enzyme in young red cells of primaquine-sensitive subjects and the deficiency shows up only among the elderly red cells which otherwise manage to carry on their appointed task. However, when the added burden of a drug like primaquine has to be taken on, these cells just give up the ghost and disintegrate. The hemolytic anemia which ensues can be quite dangerous. It became apparent that primaquine is not the only toxic agent which menaces such individuals. Naphthalene is one such toxic agent which, when brought into the home as a moth repellent, can cause a severe reaction in children with this genetic aberration which deprives their red cells of their complete endowment of enzymes.

Once we knew what enzyme deficiency to look for, another entirely different hemolytic anemia was found in which there is also a shortage of the same enzyme. Among peoples who inhabit the periphery of the Mediterranean basin there is a truly esoteric ailment. In those who are afflicted, the eating of the fava bean (*Vicia fava*) is followed by a severe bout of hemolytic anemia. The attack is particularly severe if the raw bean is eaten—by hungry children. This syndrome, to which the name favism has been assigned, is but another manifestation of the shortage of glucose-6-phosphate dehydrogenase. Whether overburdened by primaquine, naphthalene, or some unknown agent in the fava bean, the red cells of the afflicted throw in the sponge.

Favism is a hereditary inborn error of metabolism which became recognized not by the excretion of some accumulating product but by a diligent search for the cause of the toxicity of a drug. Some fascinating studies on the distribution of favism have been conducted in Israel. That country is a rich laboratory for the study of genetics among migratory but isolated peoples. As a result of various persecutions there is an influx of peoples from all over the world who thus offer rich material for study, and sufficient numbers of trained scientific personnel are on hand to accomplish the investigations. Favism is prevalent among the Sephardic Jews—those who remained around the Mediterranean area after the destruction of the Temple. But the syndrome is essentially absent among the Ashkenazic Jews—those who were scattered all over Europe.[3]

The inborn errors of metabolism which are clinically recognized are the result of gross aberrations within the genetic

---

[3] Ironically, the Yemenite Jews, who fled Yemen because of Arab hostility, are identical genetically with the Yemenite Arabs. Clearly a case of blood not being thicker than nationalist propaganda.

makeup of an individual. There are other gross differences in our genetic makeups which do not manifest themselves through any clinical symptoms. The reason is simply that the function of the enzymes involved is so peripheral that no disadvantage accrues from its loss, or gain. For example, there is a drug, phenylthio-carbamide (let us call it PTC for short), which, when placed on the tongue of a hundred humans will evoke a bitter taste in about sixty-six and absolutely no response in the others. Genetic analysis revealed that "tasting of PTC" is inherited as a Mendelian dominant trait. What that particular "tasting enzyme" is doing in the tongue of some of us we do not know, and its absence certainly cannot be classed as an inborn "error" of metabolism. There are finer fluctuations in enzyme levels among normal individuals. No two human beings are exactly alike. Differences in gross structures are obvious: identification by our fingerprints takes advantage of just one area of individuality. That we are individualistic even at the molecular level has become obvious from the work of Dr. Roger J. Williams of the University of Texas. He and his associates measure in human subjects a variety of metabolic parameters. These include components of the blood, excretion of products in the urine, and measurable reactions to certain drugs—including "tasting and nontasting" of some. From these measured entities a chart is prepared for each subject. This is in the shape of a star with lines of different length radiating from a central point. Each line represents a measured entity, the length of the line indicating the quantity. Thus a lot of information is concentrated in a small area and a sort of metabolic fingerprint is obtained for each subject. No two persons show the same pattern, indicating that we differ not only in gross structures but even at the level of our molecular components. The only exception to this rule are identical twins, whose metabolic fingerprints are always essentially the same. Since identical twins originate from one cell and are thus genetically identical, and since it is only they who have similar metabolic patterns, we have evidence of genetic control at the molecular level.

Our genes determine not only whether we are tall or short, dark or light, but also how we metabolize our sugar or cholesterol, what is the level of hormones in our blood, how we react to nicotine—in short, they are the ultimate determinants of the many-faceted jewel that is life.

5 PROTEIN STRUCTURE IN HEALTH AND DISEASE

"PRESENT-DAY biology is dominated by the enormous successes of biochemistry." This encomium is not jejune self-praise from some partisan enthusiast of my profession. It is from C. H. Waddington, an outstanding contemporary English classical biologist. (A classical biologist is easily defined by a simple exclusion. He is one who is working in biology not at the molecular level but with intact cells or whole organisms.)

Perhaps for no achievement is Waddington's endorsement more deserved than for the unraveling by the chemist of the intricate structure and wondrously varied functions of the proteins. This was a cardinal contribution to our understanding of the process of life. After all, every dynamic act a living cell can perform is done through the indispensable agency of proteins. Without enzymes—all of which are proteins—the nucleic acids are inert chemicals, their hoard of encoded information as useless as a volume of Shakespeare in the hands of an illiterate Australian bushman. For as we shall see, only enzymes can decipher the Code of Life.

The proteins are the most abundant solid components of the cell. For example, even in a bacterial cell which grows very rapidly and is therefore high in its nucleic acid content, half of the total solids are proteins; about 5 percent is DNA and 25 percent is RNA. A very small fraction of those proteins is delegated to make more DNA and RNA and the rest are pressed into service to perform the vast variety of functions the sum total of which is life. Proteins run the whole gamut of reactivity: they can be inert as the scales on a fish or as frenetically active as some enzymes which can repeat chemical reactions two million times a minute. The molecular size of proteins is equally wide-ranging. They can be as low as 6000 and as high as a million or more. Every protein molecule appears to be designed with high specificity for its assigned task. The specificity is achieved by the almost infinite variations in structure with which protein molecules can be constructed.

The building units for the shaping of a protein molecule are the amino acids. There are twenty of them and in their molecular design they represent the ultimate in diversity and unity. All but one of them have the following pattern of atoms as part of their structure:

$$R-\underset{\underset{NH_2}{|}}{\overset{\overset{H}{|}}{C}}-COOH$$

The name, amino acid, is derived from the atomic attachments on the central carbon atom. It holds an amino group (NH_2) and an acidic one (COOH). R is a shorthand symbol for an array of atoms which is different in every amino acid. It can range from a hydrogen atom in glycine to a complex double ring in trypto-phan.

$$\text{H}$$
$$|$$
$$\text{C}$$

$$\text{H}$$
$$|$$
$$\text{H—C—COOH}$$
$$|$$
$$\text{NH}_2$$

$$\text{H—C} \quad \text{C} \quad\quad \text{C—CH}_2\text{—C—COOH}$$

$$\text{H—C} \quad \text{C} \quad\quad \text{C—H} \quad \text{NH}_2$$

$$\text{C} \quad\quad \text{N}$$
$$| \quad\quad |$$
$$\text{H} \quad\quad \text{H}$$

Glycine   *Tryptophan*

The deduction of the method of attachment of amino acids to form the giant protein molecules is a splendid achievement of the German school of organic chemistry which flourished until Hitler's Third Reich. Emil Fischer and Albrecht Kossel were the redoubtable masters who at the turn of this century pried out from the protein molecule the secret of its assembly. The acidic group of one amino acid can form a firm bond with the amino group of another amino acid by the mutual shedding of a water molecule. The electronic forces which had originally held onto the atoms that form the water now graft the two amino acids together:

$$\overset{1}{\text{R}}\text{—C—C—}\boxed{\text{OH H}}\text{—N—C—C—}\boxed{\text{OH H}}\text{—N—C—COO H}$$

This accretion can continue until hundreds of amino acids are securely attached, producing molecules of gigantic size and intricate structure.

A protein molecule has a homogeneous backbone from which protrude twenty different molecular side arms. Since the central core of all proteins is essentially the same, the wondrously varied capabilities of the vast number of different proteins must be derived from the distribution of those appendages. Since there are twenty different protrusions, the variety of permutations possible is almost limitless. For example, with only three different amino acids the possibilities are six entirely different compounds. Let us take the three amino acids glycine, leucine, and alanine and list the six possibilities. (The abbreviations gly., leu., and ala. are widely used.)

gly · ala · leu
gly · leu · ala
ala · gly · leu
ala · leu · gly
leu · gly · ala
leu · ala · gly

With one of each of twenty different amino acids the number of permutations is twenty! (factorial twenty) which is equal to 2×10^{18} different protein molecules. Yet such a protein would be puny as proteins go, weighing only 2000 atomic weight units, whereas some proteins run up into the hundreds of thousands. With some of these the possible permutations become truly astronomical. For example, Dr. Synge, one of the co-discoverers of paper chromatography, did some chemical sheep counting one sleepless night. He envisioned a protein of modest size— 34,000—which contained not twenty but only twelve different amino acids. In this case the permutations of different proteins which are possible run to the unimaginable number, 10^{300}. Should only one single molecule of each of these species exist, their combined mass would be 10^{280} grams. But fortunately the total mass of our planet is only 6×10^{27} grams and therefore we can dismiss all thoughts of these monstrous numbers of proteins.

Until about twenty years ago the possibility of decoding the exact sequence of amino acids in a natural protein appeared to be a staggeringly complicated problem, beyond the reach of our best minds and tools. Take the case of one of the simplest proteins known, the hormone insulin. From the searching studies of many chemists, we knew twenty years ago that insulin contains sixteen of the known amino acids and that its molecular weight is 6000. This indicated that some of the amino acids appear in the molecule more than once. (Actually there are sufficient repetitions to add up to fifty-one amino acid units.) The different sequences possible are sufficiently frightening to make anyone seek a simpler, more sensible problem as a challenge. However, a young Englishman, Dr. Frederick Sanger, was not frightened. He had two novel tools for his vast enterprise. One he designed himself, the other was given to him by two other Englishmen, Drs. A. J. P. Martin and R. L. M. Synge.

Martin and Synge came to the rescue of every one of us who struggled in biology with chemical tools and who raged in frustration at the inadequacy of the analytical methods which could not reach down to the low levels in which many biologically important substances are present in the cell. Martin and Synge developed paper chromatography, which achieved at least a thousandfold refinement over previous methods and thus opened up recesses in the chemical mechanisms of the cell which had lain beyond our reach. The term chromatography was originally applied to the separation of a mixture of plant pigments by the Russian botanist M. Tswett. He adsorbed a mixture of plant chlorophylls on a column of powdered limestone. He found that if a solvent for the chlorophylls, petroleum ether, was poured onto such a column the components of the pigmented mixture became separated and moved downward at different rates, forming well-delineated colored bands. The rate of movement of each different pigment is governed by two attributes: the solubility of the pigment in the solvent, and the force of association of the pigment and the powdered limestone.

Paper chromatography is a variation of this principle. Martin and Synge decided to eliminate the column of powdered limestone and used instead a sheet of filter paper for rigid support. The substances to be separated are placed in a tiny drop of solution on the bottom of the filter paper. The paper is rolled into a cylinder and the lower edge of it is inserted into a liquid that usually has more than one component and the paper is left standing for several hours. The liquids will rise, defying gravity, along the thin fibers of the paper, and they will carry with them the solid components of the mixture that had been deposited on the bottom of the filter paper. But the solids do not ascend at the same rate, for the rate of the migration is an unvarying characteristic of each species of molecule. Thus the components of a mixture which had taken theretofore perhaps a year of chemical drudgery to separate now can be taken apart in a matter of hours. And once they are separated, identification is easy, for the very height (relative to that reached by the solvent) to which a substance rises on the filter paper is a characteristic identifying attribute—like a whorl in a fingerprint. For the location of a given substance on the filter paper there is a whole catalogue of chemical reagents which can yield colored products and thus define the area reached by the unknown substance. The enormous advantage offered by paper chromatography is best seen by a comparison. About twenty-five years ago a well-known English chemist achieved the first reliable assay in milk protein of one of the amino acids, glutamic acid. The study required a whole year and he needed a hundred grams of casein as his starting material. Today an assistant skilled in the techniques of paper chromatography can perform a dozen assays of this amino acid in twenty-four hours and he needs but one millionth as much starting material. Figure 5.1 is a schematic diagram of paper chromatography.

With the tool of paper chromatography in hand, Dr. Sanger was ready to tackle the most intricate cryptogram ever to challenge the human mind: the decoding of the structure of a protein. His stratagem was dictated by the unique mode of attachment of amino acids. At one end of the chain there is an amino acid whose amino (NH_2) group is unengaged; at the other end an acidic group is free. Sanger decided to develop a method to pinpoint the amino acid with the free amino group. This would, of course, reveal the first amino acid in the chain. There had been many previous attempts to hang a bell on the free amino group of the first amino acid, but success was limited at best. A chemical agent for such a task must be a molecule which seeks out only the free amino group and then adheres to it so tenaciously that the protein can be dismembered by heating with acid without removing the label. If in the debris of amino acids the one with the label can be found we can conclude that

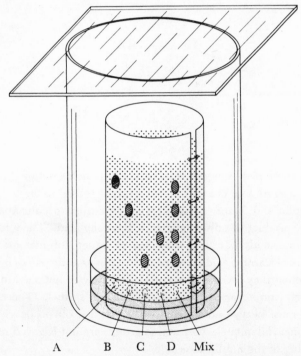

Fig. 5.1. Schematic presentation of an ascending chromatogram. The four components of the mix are tentatively identified as A, B, C, and D, since they migrate up the paper cylinder as far as the four known markers

we have the amino acid at the head of the chain. Dr. Sanger
devised a remarkably effective label for this task.

It is this compound which is usually called Sanger's reagent:

$$NO_2 \quad \begin{matrix} F \\ | \end{matrix}$$

$$NO_2$$

It combines readily with an amino acid in the following way:

$$\begin{matrix} R \\ | \\ H-C-COOH \\ | \\ N-H \end{matrix}$$

$$NO_2$$

$$+ \ HF$$

$$NO_2$$

Once combined, Sanger's reagent not only holds onto its amino
acid, it serves as a beacon: it has a bright yellow color.

Insulin and Sanger's reagent were permitted to unite and the
yellow product was dismembered with strong acid. The solution
of amino acids was then subjected to paper chromatography.
The sixteen amino acids arrayed themselves as anticipated, but on
inspection they revealed a surprise. There were not one but two
different amino acids carrying Dr. Sanger's label. Therefore,
there must be two amino acids—they turned out to be glycine
and phenylalanine—whose amino groups are not engaged in the
weaving of the insulin molecule.

This intelligence could be interpreted only one way: Insulin
is not a single strand of amino acids; it must be a double one.
How are these two strands held together? Insulin is rich in the

sulfur-containing amino acid cysteine, which has a unique attribute. It has not two but three hooks with which it can form protein links. In addition to its amino and acidic groups its sulfur atom can reach out to the sulfur atom of another cysteine in a neighboring strand and forge a fairly strong link.

Our knowledge up to this point can be summarized with the diagram below:

A   *Glycine* ———————— *Cysteine*
                          |
                          S
                          |
                          S
                          |
B   *Phenylalanine* ———————— *Cysteine*

The reader should not feel that he is being patronized with the use of this simplified symbolism. All chemical formulas are crude symbols which summarize the sum total of known attributes of a substance. The formula

$$NO_2 - \overset{F}{\underset{NO_2}{\bigcirc}}$$

no more depicts the real physical structure of that molecule—with its scores of subatomic particles whirling around—than does a dot on a map reveal the sum total of New York City.

The next order of business was to separate the two chains, A and B. This was relatively simple, since the sulfur links were known to be easily broken by oxidation, but they had to be separated. At this point Dr. Sanger's ingenuity was matched by good luck: the two chains turned out to have different electrical charges on them. Molecules with different charges migrate in different directions in an appropriate electrical field, and we are therefore able to separate such a mixture into two homogeneous components.

The two different chains produced by the cleavage of the sulfur bridges were separated and purified. Chain A, which had glycine as its first amino acid, had about twenty amino acids in it; chain B, whose guide bearer was phenylalanine, had about thirty.

Chain B was tackled first because it promised to be easier. There is a larger variety of amino acids in it, therefore there was a likelihood that some of them might appear only once. In chain A, with a smaller variety of amino acids, the probability of repeated appearance is greater, which places an extra burden on the decoding.

Chain B was treated with Sanger's reagent and was subjected to very mild fragmentation. This is done with weak acid and very gentle heating. Under these conditions the cleavage is incomplete, yielding fragments of the chain containing two, three, and four amino acids still hooked together.

Such a mixture of debris was paper chromatographed and allowed to migrate according to the propensities of the components. (Their rate of travel is slower than that of individual amino acids.) Dr. Sanger observed several different yellow spots. Any labeled phenylalanine that was free of all other amino acids formed one spot. A phenylalanine that still had the next amino acid attached to it formed another spot, one with two amino acids attached to it still another, and so on.

Each of these spots on the paper was cut out, the amino acid chains were leached out of them, and each solution was evaporated. Dr. Sanger now had a series of fragments of the original chain.

Each of the leached-out spots was cooked with acid to break them down completely, and they were analyzed by paper chromatography for the identity of the amino acids.

Dr. Sanger was successful in finding fragments which contained only two amino acids, the labeled phenylananine plus another amino acid. It turned out to be valine. Now the decoding was under way. The first two amino acids of chain B are: phenyl-alanine-valine. A triplet of amino acids was also found which contained labeled phenylalanine, valine, and a third amino acid, aspartic acid. Therefore, the sequence of the first three amino acids must be

Phenylalanine—Valine—Aspartic Acid
   1   2   3

A quadruplet of amino acids revealed the sequence:

Phenylalanine—Valine—Aspartic Acid—Glutamic Acid
   1   2   3   4

A quintuplet of amino acids yielded the following sequence:

Phenylalanine—Valine—Aspartic Acid—Glutamic Acid—Histidine
   1   2   3   4   5

This was the longest of the chains still containing labeled phenylalanine that Dr. Sanger could find. At this point he resorted to other tactics. He turned to the amino acid complexes which did not have the labeled number 1 attached to them. This involved examination of the mixtures which originated from further along the chain. He found a triplet which contained amino acids 4 and 5 and a new one which was not 3. A chain of thirty amino acids can form only twenty-eight triplets. Since there are sixteen different amino acids to be distributed among those twenty-eight triplets, the probability of both 4 and 5 appearing more than once is small. Therefore the triplet containing 4 and 5 and the new one most likely was 4–5–6. Amino acid 7 was found by obtaining a triplet containing 5–6 and a new one.

Dr. Sanger continued hunting and matching overlapping fragments with the sensitivity of an artist, the concentrated intensity of a genius, and the ingenuity of a truly original mind.

The sequence of some of the fragments from further along the chain could be decoded only by labeling their first amino acids with Sanger's reagent and performing a stepwise fragmentation of them. As the sequence was laboriously wrested from each fragment it was fitted into the larger pattern until, finally, thirty amino acids were unequivocally delineated.

He then tackled chain A and, using the same strategy, decoded a sequence of twenty-one amino acids.

He next attacked the question of the location of the sulfur bridges through which the two chains are connected. It is easier to visualize his approach if we line up the two chains side by side, designating the amino acids by numbers (see following page). In chain A cysteine appears four times: 36–37, 41, 50. In chain B it makes its appearance only two times: 7, 19.

The question is which of the six sulfur atoms are the abutments through which the bridge or bridges are formed. To answer this question Dr. Sanger's ingenuity once again provided a unique approach. He subjected whole insulin, with the sulfur bridges intact, to disintegration by enzymes which were known not to sever sulfur bridges, and he started once again to hunt for fragments. This time he sought only fragments which contained cysteine. He found novel sequences which exist neither in chain A nor B. For example, he found a fragment of four amino acids, such as this:

$$\overset{8}{\text{Glycine}}—\overset{7}{\text{Cysteine}}—\overset{37}{\text{Cysteine}}—\overset{38}{\text{Alanine}}$$

Since such a sequence appears in neither chain A nor B, the two cysteines must come from the two different chains carrying a neighbor with them; therefore, there must be a bridge between 37 of A and 7 of B.

Similarly, he found another bridge between 59 of A and 19 of B.

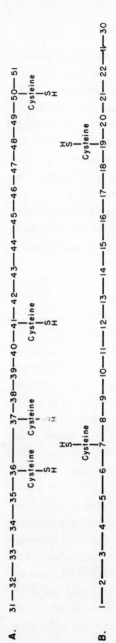

A. 31—32—33—34—35—36—37—38—39—40—41—42—43—44—45—46—47—48—49—50—51

Cysteine Cysteine Cysteine Cysteine
—S —S —S —S
H H H H

B. 1—2—3—4—5—6—7—8—9—10—11—12—13—14—15—16—17—18—19—20—21—22—...—30

H H
S S
Cysteine Cysteine

The location of the sulfur-bearing cysteines

A. NH₂S—
|
NH₂ NH₂ NH₂ NH₂
| | | |
Gly·Ileu·Val·Glu·Glu·Cy·Cy·Ala·Ser·Val·Cy·Ser·Leu·Tyr·Glu·Leu·Glu·Asp·Tyr·Cy·Asp
S— —S
| |
S—— —S

B. NH₂ NH₂
| |
Phe·Val·Asp·Glu·His·Leu·Cy·Gly·Ser·His·Leu·Val·Glu·Ala·Leu·Tyr·Leu·Val·Cy·Gly·Glu·Arg·Gly·Phe·Phe·Tyr·Thr·Pro·Lys·Ala

The structure of insulin

Thus, after a decade of intense dedication, Dr. Sanger could unfurl one of the proudest pennants signaling a human achievement, the structure of a physiologically active protein.

The sequence of amino acids in insulin reveals no pattern of any kind; it appears to be just one of the trillions of random sequences possible. Yet somewhere in that sequence resides the physiological potency of this hormone. Where, exactly, we do not know. Part of the chain can be altered without impairing the function of the hormone. Dr. Sanger's original work was done on beef insulin. But as the sequence of amino acids was determined in other insulins a surprise was in store for us. Chain B was the same in each one examined, but in chain A the following variations emerged:

Amino Acids

Animal	38	39	40
Cattle	alanine	serine	valine
Pig	threonine	serine	isoleucine
Sheep	alanine	glycine	valine
Horse	threonine	glycine	isoleucine
Sperm Whale	threonine	serine	isoleucine
Sei Whale	alanine	serine	threonine

These amino acids certainly cannot have unique roles within the structure of insulin. They appear to serve merely as struts to hold the structure to a definite dimension and shape. The variability of amino acids in the insulins of different orders of mammals—indeed between two different families of whales—is interesting from the point of view of evolution. Nature paints the image of life with bold uniform strokes; however, the loose bristles on the edge of her brush do often trace minor patterns distinctive of each species. No mammal has survived a mutational change involving the total loss or even gross changes in the structure of insulin. Even the changes in the three permutating amino acids 38, 39, 40 are relatively minor: the amino acids involved are quite akin in size and other physical attributes. However, all amino acid replacements are not quite so benign.

As we shall see before this chapter is finished, in some cases the interchange of one amino acid in the hemoglobin molecule of the human can spell the difference between life and death, literally.

Several different chemical forces determine the geometric form of a protein molecule. In the first place there is the permutation of alignment of amino acids which determines the sequence of side groups that stick out of the backbone of a protein. Since the core of the protein molecule—the repeating links of the chain—is identical in all proteins, the characteristic attributes of different species of proteins must be derived from the ordered array of the side groups. We call this the primary structure of the protein molecule. As we have seen, in the structure of insulin the amino acids along the chain can form liaisons on their own initiative. Two cysteines within easy atomic reach of each other can form a loop of atoms tied together by the two reactive sulfur atoms. Or, cysteines from two different chains can reach out and build a bridge and thus fuse the two strands of amino acids into a more rigid, ladderlike structure. This is called the the secondary structure of the protein molecule.

In addition, there are strong chemical forces which orient proteins into three-dimensional shapes. This architectural effect was deduced by one of the most versatile of living American scientists, Linus Pauling. Pauling explained certain regularities revealed by X-ray analysis of proteins as stemming from a repeated helical arrangement of the core of the protein structure. He pictured the atoms which form the links between the amino acids as not stretched out like a limp string but as coiled into long spirals. Forces between atoms in adjacent layers of the spiral provide added strength and rigidity and form the so-called tertiary structure. In turn, the coils of the protein molecule are intertwined to form the final edifice within which may reside the wondrous potency of an enzyme or of a hormone. Both the structure and the function of each protein are highly specific.

The determination of amino acid sequence is an onerous, time-consuming task and, therefore, very few complete sequences are known, but it is already apparent that the amino acids of different proteins are strung together in an infinite variety. Figure 5.2 represents the amino acid sequence of an enzyme, ribonuclease, which has the task and capacity to dismember RNA. Somewhere within the sequence of amino acids in insulin and in ribonuclease resides their different catalytic potency. Just where, we do not know as yet. But at the present rate of progress this secret of life's machinery cannot remain hidden for too long. For example, it is known that the amino acid histidine

$$H-C \!\!=\!\!=\!\! C-CH_2-\overset{\overset{\displaystyle H}{|}}{\underset{\underset{\displaystyle NH_2}{|}}{C}}-COOH$$

in position 119 of the ribonuclease molecule is far more active in certain chemical reactions than are the other three histidines in the molecule. That particular histidine may be the "active site" of this enzyme, the star of the show as it were, and the rest of the amino acids are merely the supporting cast.

What is the consequence of a gene mutation on the structure of the protein which stems from the altered gene? If the mutational damage is extensive the protein produced may be sufficiently altered to lose its functional capacity completely. Such an altered or incomplete protein would sit in its assigned spot on the cell's assembly line as useless as an armless man on an assembly line in Detroit. If the function of the protein is critical to the cell's functioning there is no problem, the individual in which the mutation expressed itself dies and the particular mutation vanishes. However, if the mutational change is slight there may be only minimal or no impairment in the functioning

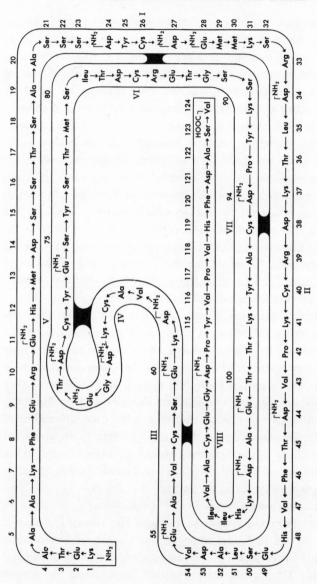

Fig. 5.2.   The amino acid sequence in the enzyme, ribonuclease

of the protein. Sanger was the first to show such benign alterations in the amino acid sequence in the B chain of insulin.
Since none of these animals are chronic diabetics these amino
acid interchanges apparently do no damage to the efficacy of
insulin as a hormone. Apparently these three amino acids are
not the active site of insulin; they may merely serve as struts to
give the whole structure its appropriate dimension. Indeed,
under the circumstances it is impossible to tell which is the
original structure of insulin and which are the mutations.

However, there are other cases where we do know the valid
sequence in a protein and its aberrant modifications. Sickle-cell
anemia is a hereditary disease which expresses itself by the
presence of crumpled, sickle-shaped red cells in the veins of the
subject. The geographical distribution of the disease is curious;
it is very high, 15 percent or over, among the populations of
central Africa and southern India. It is somewhat less frequent
among peoples living around the Mediterranean basin. In our
own Negro population about 8 percent have some sickle cells
in their veins. Fortunately only about 0.2 percent have the
abnormal cells in sufficiently large concentrations to incapacitate
the patients. The puzzling feature of the disease until recently
was that the corpuscles have their normal disc shape while they
are in the arteries, but crumple into shapeless bags in the veins.
The disease, therefore, is not caused merely by a malformation of
the cell membranes but, rather, must be due to an abnormal
response either by the membrane or its contents to changes in the
concentration of oxygen or of carbon dioxide as the corpuscle
exchanges these gases in the lungs.

Dr. Linus Pauling, a physical chemist, became interested in
biological problems several years ago. He heard of sickle-cell
anemia for the first time while he was serving on a committee
appointed by President Roosevelt to study means of advancing
medicine. Dr. Pauling very naturally approached the disease
as a physicochemical problem. He and his associates isolated hemoglobin from the blood of patients with sickle-cell

anemia. They then extracted from the hemeglobin the red pigment, heme, in a pure form and compared it with heme obtained from normal humans. The hemes from the two sources were exactly the same. The protein fraction, the globin, was next purified and was made the target of the battery of critical tests and measurements that reveal differences in proteins. One such test is the measurement of the rate of migration of a protein toward one of the poles in an electrical field. Proteins have the ability to assume either a positive or a negative surface charge depending upon the acidity of the liquid in which they are suspended. (Their ambivalence has earned them the name "zwitter-ions," meaning hermaphroditic, charged particles.) The unbalanced, excess electrical charges on the surface of the protein molecule will determine the direction of migration in an electrical field. A level of acidity could be reached at which the globin from normal red cells moved toward the positive pole and the globin from sickling cells moved toward the negative pole. In other words, the protein from the diseased cells seems to carry a larger positive charge than the normal protein does. Such an abnormal charge can have profound effects on the protein's ability to bind water or the gases carbon dioxide and oxygen to it, and, also, on the ability of the protein molecules themselves to be bound together. Pauling's discovery revealed the cause of the sickling phenomenon: When the abnormal protein is laden with carbon dioxide in the veins it shrivels in volume and fails to fill out the cell membrane. We even know the molecular site of the protein's abnormality. Dr. Vernon Ingram of Cambridge University addressed himself to this task about ten years ago. He confidently expected that the abnormal protein might have some aberration in its amino acid sequence. A complete sequence study as performed by Sanger loomed as a forbidding task: There are close to 600 amino acid units in a complete hemoglobin molecule. Ingram gambled on a modification of Sanger's approach. He fragmented the hemoglobin molecule with an enzyme—trypsin—which is known to cleave proteins only at restricted sites. Trypsin is a specialist in snipping bonds formed only by the amino acids lysine and arginine. It, therefore, can produce only chunks of the protein whose size would be determined by the number of amino acids between these two amino acids. The mixture of fragments obtained from normal and sickling hemoglobins was permitted to migrate in an electric field in one direction and was then chromatographed at right angles to their original path. The papers were sprayed with a reagent which yields purple areas where it comes in contact with amino acids or protein fragments. There were nearly thirty spots on each paper and their geographical distribution was almost identical. But there was one difference in the "fingerprints" of the protein fragments from the two sources. One fragment— arbitrarily designated as number 4—from the abnormal hemoglobin had slightly different mobility on the paper than its counterpart from the normal hemoglobin. Fragment 4 was found to contain a sequence of eight amino acids which were identical in the two samples except for one amino acid. In the abnormal hemoglobin a glutamic acid was replaced by a valine.

$$
\begin{array}{cc}
\begin{array}{c}
\text{COOH} \\
|\\
\text{HC--NH}_2 \\
|\\
\text{CH}_2 \\
|\\
\text{CH}_2 \\
|\\
\text{COOH}
\end{array}
&
\begin{array}{c}
\text{COOH} \\
|\\
\text{H--C--NH}_2 \\
|\\
\text{C--H} \\
/\ \backslash \\
\text{CH}_3\ \ \text{CH}_3
\end{array}
\\
\textit{Glutamic Acid} & \textit{Valine}
\end{array}
$$

It is this single substitution that makes the difference between healthy and crumpled red cells. Glutamic acid has an extra acidic group in its structure which is negatively charged at the levels of acidity that prevail in a red cell. This extra negative charge is lacking from the hemoglobin of a patient with sickle-cell anemia. A dearth of but one negative charge can doom a subject to the life of an invalid, and in those cases where both parents contributed a sickling gene—i.e., if the patient is homozygous—early death may be his fate.

That such a fumbling can occur in the weaving of the tapestry of the hemoglobin molecule is a source of terror; that it occurs so infrequently is a source of awe. The precision required for the perpetuation of life is almost beyond belief. A human being has thousands of different species of protein molecules. Each one of these has a unique structure; each has an amino acid sequence all its own. Billions of each of these protein molecules must be made every second of our life to shape and to sustain it. The pattern of each protein must be unerringly duplicated. If there is but one fumble and just one amino acid is misplaced, as in sickle-cell anemia, the red cells crumple shapelessly, causing disease or death. This one blunder, with its patently gruesome consequence, testifies to the unfailing success in the shaping of thousands of other kinds of protein molecules in our bodies. Disease or health, indeed, life or death, hangs on a thin thread no stronger than the link between two appropriate amino acids.

It is awesome to contemplate the extent of damage wrought in an individual by the substitution of but one amino acid. The clinical symptoms accumulate from two different stresses. The body tries to eliminate the sickle red cells by destroying them. This imposes an enormous burden of new synthesis. There is an overactivity of the bone marrow, the heart dilates, and since so much biological construction is a total waste, physical development of the individual is poor and weakness and lassitude are his lot. At the same time, the sickling cells tend to clump and thus interfere with circulation, causing a medical-textbook-full of syndromes ranging from brain to kidney damage.

How could a gene which dooms its possessor to inferior physical structure and early death survive for thousands of years? For, assuming that it was a single mutation, from the currently wide distribution of the deleterious gene the mutation must have occurred long ago. Why was it not erased from the pool of human genes by the elimination of its hapless possessors?

At first glance the sickle-cell syndrome defies Darwin's generalization on the survival of the fittest. However, this genetic paradox was resolved recently. In some areas of the earth the sickling syndrome actually confers an advantage on the person it afflicts! Sickle-cell anemia has the highest frequency in Africa and India, where falciparum malaria is endemic. It was discovered that the protozoan *Plasmodium falciparum*, which is the etiologic agent of this disease, declines to infest sickling red cells. Sickle-cell anemia thus confers immunity against malaria. This is a rare case of a deleterious mutation becoming an advantage in a hostile environment.

At the present time, even though the failure in sickle-cell anemia is clearly identified, treatment can be at best palliative. In another disease of genetic origin, however, intimate knowledge of the mechanism of the genetic failure brought complete relief from devastation by that particular scourge. Galactosemia is a rare hereditary disease whose symptoms become apparent within a few days after birth. The infant loses weight, vomits, becomes desiccated; his liver enlarges, and in severe cases, unless the disease is diagnosed, the patient is lost. Chemical analysis reveals in the blood of the subject the presence of large amounts of the sugar galactose. Hence, the name galactosemia. Galactose makes up half of the structure of the sugar of milk; lactose,

Galactose   *Glucose*

Lactose

the other half, is made of grape sugar, or glucose. The only difference between the two sugars is the geometric pattern of the atoms attached to the fourth carbon atom. The H and OH are located on opposite sides of the carbon atom in the two sugars. The enzymes whose task it is to metabolize sugars are highly specific. They are tooled to the metabolism of glucose and not of galactose. Lactose brings to the infant an oversupply of galactose.[1]

The oversupply of this sugar is managed in a normal infant by an enzyme which attaches itself to galactose and swings around the appropriate atoms as the wind flips a weathervane. This sleight of enzyme transmutes the galactose into glucose and the normal metabolism of that sugar can then proceed. Very rarely, through a genetic blunder, an infant is born without the capacity to make enough of the atom-shuffling enzyme. The results of this deficiency of one enzyme were ghastly until a few years ago. Every tissue of the afflicted would be flooded with the unused galactose. Essentially, a crippling, esoteric diabetic condition would set in. The kidneys would attempt to cope with the emergency by excreting the galactose but they would be overwhelmed. The infant would either succumb or at best, or worst, would survive with permanent damage to its organs, including the brain. Today, provided diagnosis is early enough, we can rescue the child by an astonishingly simple expedient: since milk is the only source of galactose we exclude milk from the infant's diet. One component of the diet, or the lack of one enzyme for its metabolism, spells life or death. We could intercede in this case with such ease because we understand the syndrome and its cause at the molecular level. For a long time to come we shall be

---

[1] It is interesting that lactose is not found anywhere in nature except in milk. The mutation which produced the enzyme for the synthesis of lactose must have coincided with the mutations which gave rise to the development of the mammary gland.

unable to alter the genetic mechanisms to restore a functional enzyme. However, the remedial step which we can take in galactosemia is just as effective. We shall find other therapeutic bypasses of other diseases as we master molecular mechanisms in health and disease.

6 THE GENE IDENTIFIED
DNA, THE MASTER MOLECULE

RESEARCH institutes, like dynasties, flourish, soar to heights, become complacent, and languish. The Rockefeller Institute in New York was undoubtedly the outstanding center of research in the biological sciences in our country in the first quarter of this century. This excellence was the result of the fruitful coincidence of adequate funds, creative professional leadership, and the scarcity of other research institutions. It must be recalled that prior to 1940 we Americans spent twice as much on funeral flowers as on medical research, and consequently in the second and third decades of this century the Rockefeller Institute could have its pick of young scientists complete with Ph.D. or M.D. and European postdoctoral training for $1,000 a year. The picking was done by the head of the institute, Simon Flexner himself, who proved to be an extraordinarily astute talent scout.

The man who was to determine finally the chemical nature of the gene, Oswald T. Avery, was a worker at the institute during its golden era. Avery went into research as a refuge from the practice of medicine. A small, shy, sensitive man, he suffered

Fig. 6.1. Oswald Theodore Avery, M.D. (Columbia University), 1877–1955

doubly with each patient: from the sickness itself and from the frustration at the limited arsenal he had against the disease.

Once he was given the opportunity in 1913 to devote his time to research he made the most of it. He let nothing interfere with his work. He lived across the street from the institute, he never married, and he hardly ever traveled. He husbanded his time and energy to the point where he would not answer correspondence notifying him of the receipt of some award. For the Professor, or "Fess" as he was affectionately called by his assistants, gained recognition for his studies of the etiological agent of pneumonia. It was probably his medical experience which induced him to work on the pneumococcus, which in those preantibiotic days was the number-one killer of man.

Avery worked with only two or three associates at a time, several of whom were to achieve preeminence in their field. He published but rarely, only when he felt the contribution was significant. Those were calm and happy days for a real scientist. There were no grants, therefore neither grant applications nor progress reports to write. The rewards for the scientist in money, social prestige, and power were practically nil. The only rewards were the joy of work and the appreciation and esteem from one's peers.

Avery attacked the problem of pneumococcal infection on a broad front. With an associate of his, Dr. A. R. Dochez, he undertook a systematic study of the immunological properties of the pneumococcus. From these studies emerged not just information about this one pathogen but the foundation structure for a new branch of biological science, immunochemistry. The architect of the new structure was a young chemist at the institute, Dr. Michael Heidelberger.

Avery and Dochez found in 1917 a brand new tool for immunological studies when they discovered that virulent forms of pneumococci secreted into the culture medium what they called a "soluble specific substance." The soluble substance could act as a surrogate for the whole bacterial cell in immunological reactions. If a culture of virulent pneumococci is killed by steam and is injected into a rabbit, nothing unusual happens. But, if in a week or two we inject the same rabbit with an identical dose of the dead bacteria there will be a violent reaction. The animal's breathing will become labored, it will thrash around, and will finally go into shock, from which—depending on the dose injected—it may not recover. The rabbit will have an altered reaction or an allergy to the second injection.

This violent response to the second injection is a part of the body's defense mechanism. Animals resist the entry of foreign proteins into their tissues. The proteins we eat do not normally enter our tissues intact. The proteins are broken down by the enzymes of the alimentary canal into their component amino acids and only these are permitted to enter our tissues. From the absorbed amino acids we fashion protein molecules in the image of our own proteins. In patients who suffer from the various allergies there seems to be a minute amount of seepage of the foreign proteins into their tissues. Therefore, if a foreign protein does enter into the tissues of an animal it means that a stranger is within the gates. The stranger may be just a protein molecule or a whole organism with its foreign proteins. The reaction of the animal is the same to either danger. It begins to fashion shock troops, or antibodies, in an attempt to dispose of the invaders.

The antibodies form, with great specificity, insoluble precipitates with the antigen, which may be a whole organism or a protein from it. Others before Avery had also observed that extracts of bacteria could serve in lieu of the whole organisms in eliciting immunological reactions in animals, but those investigators lacked the intuition, the self-confidence, and the persistence to seize the opportunity on which, as we shall see, the structure of a new science could be erected.

Avery recognized that his "soluble specific substance," in which the antigenic attribute of the pneumococcus may be invested, should be explored and its trail followed wherever it may lead. He wanted to learn some of the chemistry of this substance but felt that he lacked the training for such a task, so he kept needling Dr. Heidelberger to tackle it.

"The whole secret of bacterial specificity is in this substance," Avery would say as he showed Heidelberger some dark-brown paste. "When are you going to work on it? When will we find out the chemistry of it?"

When Heidelberger could put aside other duties, he began, in 1922, a painstaking, meticulous purification of the type specific substance. His perfectionist skill was rewarded. He found no nitrogen in the purified preparations and therefore the "specific substance" could not be a protein. Heidelberger was thus the first to discover that substances other than proteins can serve as antigens and evoke an immunological response. The "specific substance" proved to be complex sugars from the bacterial capsule.

Heidelberger seized the opportunity presented by a nitrogen-free antigen to study the chemistry of the interaction of antigen and antibody. At that time this field was in chaotic disarray. What the field needed to become an exact science were, in Heidelberger's words, "exact analytical chemical methods of measurement so that one could tell in absolute units, that is in units of weight, how much antibody an antiserum contained."

He was able to design such methods because he had antigens which were nitrogen-free sugars and, therefore, when antigen and antiserum were mixed, all the nitrogen of the precipitate came from the antibody. Hence, a simple assay for nitrogen could determine the total amount of antibody. With patience and high technical skill Heidelberger kept purifying antibody prep-arations until he obtained samples which were completely precipitable with antigens. That is, his preparations contained no proteins other than the specific antibody to the complex sugar of the pneumococcus. Since it was a pure protein it could be characterized by the battery of tests which by then had become perfected. The antibody turned out to have the attributes of a group of proteins of the blood serum which are called the globulins. Depending on their rate of migration in an electrical field, the globulins had been separated into various fractions by Dr. Arne Tiselius of the University of Uppsala. The antibody which Dr. Heidelberger had isolated in pure form had the characteristic mobility of a fraction of blood proteins which Dr. Tiselius had arbitrarily designated with the Greek letter γ (gamma). Thus, for the first time, we knew in what fraction of the blood's many proteins the antibodies were to be found, and gamma-globulin preparations for reinforcing a subject's own pool of antibodies became an added weapon in our arsenal against disease.

From Dr. Heidelberger's studies it became apparent that the particular strain of pneumococci he was using was not unique in producing a specific complex sugar which could elicit the immunological reaction. Literally scores of different pneumococci were found which produced complex sugars, each of which could provoke an animal to produce an antibody to it. In turn, the many proteins of each pneumococcus could also challenge the host animal to antibody production and thus Avery's search for a single component of bacteria that gave their immunological specificity bogged down in the awesome complexity of biological mechanisms.

However, even though Avery was temporarily barred from his original goal, his efforts of this period were by no means fruitless, for with uncanny intuition he chose the tools with which Dr. Heidelberger could erect the foundations of a new scientific discipline, immunochemistry.

Someone else soon picked up a new experimental trail in the search for bacterial specificity. This was the right path, for it led directly to the ultimate determinant of individuality of all things living; it led to the gene itself.

Dr. F. Griffith was a medical officer in the British Ministry of Health who, almost forty years ago, was studying the difference in virulence of various types of pneumococci. He isolated single bacterial colonies from the sputum of patients with lobar pneumonia, grew pure cultures of the organisms, and studied their virulence by injecting such homogeneous bacteria into mice.

He noted that the pneumococci which, on special nutrient plates, form little mounds of cells that have a rough surface were harmless when injected into mice and those that form smooth, glistening mucoid mounds were virulent. To study whether a harmless culture may revert to virulence he killed a batch of smooth, virulent organisms with steam and injected this preparation along with live, rough, attenuated organisms into mice. The concoction proved to be lethal. When Griffith performed a bacteriological post-mortem he made an extraordinary observation. He found that the agents of death were live, smooth pneumococci which were teeming in the blood of the dead mice.

In the animal, the rough, harmless cells were transformed into smooth, virulent ones. Griffith's explanation of his finding in 1928 may have been influenced by the preoccupation of scientists of that period with nutritional problems. He thought that the dead, smooth pneumococci broke up in the body of the mouse and furnished a "pabulum" which the live, rough ones utilized to build up a smooth structure.

But the molecular mechanism of this spectacular transformation in structure and virulence proved to be far more sophisticated than could be visualized by its discoverer. Griffith had actually observed a gene in action!

However poorly Griffith interpreted his findings, he should be given full credit for describing a completely new phenomenon of profound potential significance. It is not easy to dig in the garden of biology where thousands of scientists have already wielded their tools and to unearth something truly new. Developing and refining phenomena previously recognized by others is relatively easy: everyone who has to eke out a career in science can and does manage just that. But to observe and lift out something really new requires special gifts. It requires a humility before nature and a hauteur toward one's colleagues. Nature's clues must be revered, prevailing dogma rejected. The scientist who can pioneer must have insight, he must know how to design a clear-cut experiment; he must have sufficient self-confidence to be sure that what he observes is real, and sufficient mastery of the field to know that what he has found is new. And, finally, a dash of luck helps too.

It is not too surprising that the first step toward the chemical identification of the gene should have been taken not by a biochemist but by someone who was a biologist. Biological experiments are much easier than biochemical ones for the simple reason that it is the intact organism which completes the experiment after the initial challenge.

Griffith's finding came to the attention of Dr. Avery, whose first reaction was incredulity, for what Griffith was claiming was nothing less than transforming the organism from producing one type of complex sugar of the bacterial surface to another type. The "smooth" and "rough" textures are but the visual expressions of the kind of complex sugars the pneumococci elaborate (see Figure 6.2).

Dr. Avery asked one of his young associates, M. H. Dawson, to try to repeat the work. Repetition of a phenomenon reported by someone else is always the first step in any program of exploration or expansion of that finding. It is considered to be the

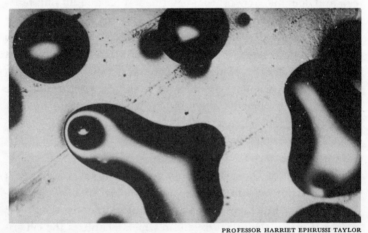

Fig. 6.2. The upper panel is a photograph of "rough" pneumococci. The lower one is that of "smooth" organisms.

responsibility of the original discoverer to describe his finding and his methods with sufficient detail and accuracy to enable anyone with normal skills in the field to repeat the experiments successfully. The methods, the terminology, and the units of measurement in the physical sciences are so universally uniform that repetition of reported experiments is successful in the vast majority of cases.

Dawson not only repeated Griffith's experiment, he and an associate of his extended it. Dawson and Sia added to a test tube culture of rough cells two preparations: heat-killed smooth cells, and serum from a rabbit that had been immunized against rough cells. The strategy of their experiments was simple. The rabbit serum would inhibit the growth of the rough cells and thus those bacteria which were transformed to smoothness would have a selective advantage in growth and cell division. Their scheme worked, and they thus proved that the mouse was not an obligatory participant in transformations: a test tube could substitute for it.

Four years after Griffith's description of his experiment J. L. Alloway, who was also at the Rockefeller Institute, took a big step forward in understanding transformation. He paved the way for a chemical exploration of the nature of the agent that is responsible for transformation. Up to this point everyone used heat-killed donor cells that were not disintegrated. It was uncertain, therefore, whether the transformation was achieved by some organized structure of the cell or by some soluble smaller component. Alloway prepared extracts of the donor cells and passed these extracts through a filter which eliminated all debris and organized elements as well as any surviving cells. The clear, sterile solution which Alloway's painstaking efforts produced still retained the transforming potency.

As Tennyson put it, "Science moves but slowly, slowly, creeping on from point to point." Twelve years were to elapse between Alloway's demonstration that some soluble substance was the transforming principle and the eventual identification of that agent.

At this point Avery himself took over the task of determining the chemical nature of the transforming principle. It is characteristic of many great biological scientists that at the full tide of their careers they can become patient students of bordering disciplines if their biological studies force them to wield the tools of those neighboring crafts. By the early 1940s Avery did not feel the need, as he did two decades before, to plead with a chemist to tackle a biochemical problem. He was by then in full tide of self-confidence as a researcher. His own work and the efforts of his former associates whom he had guided to frontiers rich and yielding were blossoming. He was surrounded by young associates who were eager to work. He exhorted them to design clean-cut experiments. "The best experiment," he would say, "needs but two test tubes: one that works and one that doesn't." He generously shared his mature knowledge and synthesis of scientific literature. When in the mood, he would hold extensive discourses on a variety of biological problems. The assistants would irreverently refer to these lengthy monologues as "red discs," i.e., phonograph records. The relationship between a master scientist and his younger associates is a complex one. It does not have the element of idolatry of the relationship of, say, a master musician and his school of younger followers. The young scientist, if he is any good, is chafing to do something new and different and soon resents the domination of the experienced and successful guide. Such mild resentment must be at the root of the deprecating allusion to Avery's discourses as "red discs" by his younger colleagues.

He undertook the search for the transforming principle with the aid of two young research physicians, Colin M. Macleod and Maclyn McCarty. In a scientific communication which has become a classic, they stated that their goal in the study of the transformation phenomenon was "to isolate the active principle from crude bacterial extracts and to identify, if possible, its chemical nature or at least to characterize it sufficiently to place it in a general group of known chemical substances."

Avery's contribution is a shining example of science at its best: the work was painstakingly thorough; the methods well defined and repeatable; the interpretation was modest yet imaginatively integrated into the other large currents of biological thinking.

They chose for their studies the transformation of a rough pneumococcus Type II to a smooth, Type III, i.e., one with a glistening mucoid coat. The designations Type II and Type III are descriptions of the bacteria's antigenic attributes. As a result of Avery's initial observation and of Heidelberger's meticulous studies, workers in the field no longer needed to depend solely on the visual differences among the pneumococci, but could also distinguish them by the exquisitely discerning—and quantitative —tool of immunochemistry. Serum of a rabbit sensitized by Type III pneumococci will give a positive reaction—a chemically measurable precipitate—only with Type III organisms. Therefore, the efficacy of the transformation could be assessed not only by visual examination of the appearance of the bacterial clones, but by a quantitative determination of the precipitation of the erstwhile Type II bacteria by Type III specific antisera.

First of all they addressed themselves to the standardization of the biological system, for they found the transformation process to be highly temperamental. Not only had the extract of the donor cells to be perfect, but also the recipient pneumococci seemed to have phases in their growth when they were more amenable to transformation.

The mechanism of this variability in receptivity—or competence as it came to be known—became a separate area of study. So complex is the biological world, and therefore so slow is progress in its exploration, it has taken twenty years to begin to understand the mechanism of competence. A former associate of Avery's, Dr. Rollin Hotchkiss, and one of the latter's young collaborators, Dr. Alex Tomasz, have recently isolated a protein that can render a recalcitrant recipient cell amenable to transformation.

Once Avery and his young colleagues had the biological system under control, they attacked the problem of the isolation of the active transforming principle. Type III pneumococci (the smooth strain) were grown in fifty-liter batches and the bacteria were separated from the fluid by centrifugation. The bacterial paste was quickly warmed to 65° and kept there for thirty minutes. This crucial step, which was essential for obtaining consistently active preparations, was devised by Alloway. He sought to destroy the enzymes which are released from the disintegrating bacteria. (Today we know that these enzymes, which inactivate the transforming preparations, specifically cleave the DNA into smaller fragments.)

In order to concentrate the active principle Avery tried exploratory forays into its possible chemical nature. This was tedious, plodding work. For example, to determine whether the transforming factor is a protein, an active preparation would be exposed to purified enzymes that were known to disintegrate proteins. The transforming capacity after exposure to the enzymes was then compared with the initial activity. Avery was greatly helped at this stage by gifts of pure enzymes from two of his colleagues of the Rockefeller Institute, Drs. Northrop and Kunitz, who had perfected methods of purification of enzymes a decade earlier to the point where the enzymes would crystallize in pure form. Had Avery used crude preparations of protein-splitting enzymes, his conclusions might have been compromised by the presence of DNA-splitting enzymes.

The transforming factor resisted the specific enzymes: therefore, it was not a protein. This came as a great surprise and a great advantage. It had been known for almost two decades that all enzymes are proteins, and until Avery's experiment no component of a cell other than a protein could be induced to function when removed from the organized structure of the cell.

That a potent factor which achieved transformation should not be a protein was totally unexpected. On the other hand, since almost half of the total solids in a bacterial cell tend to be proteins, it was an asset for the purification procedure to get rid of so large a fraction of inert solids. The elimination of protein is achieved by shaking the aqueous solution of it with chloroform. This chemical manipulation tends to distort the molecular shape of proteins so they become insoluble in water and they form a semisolid foam with the chloroform. The transforming principle remained dissolved in the water and was still biologically active.

Avery also had determined that the active principle was not the complex sugar which gave the smooth cells their characteristic appearance. This complex sugar—Avery's "soluble specific substance"—was the end product of the transforming factor, not the factor itself. The complex sugar was degraded by an enzyme preparation which had been perfected by another of Avery's young associates, Dr. René Dubos. (Dr. Dubos was to be the first man to isolate an effective antibiotic from microorganisms.) But now, the solution of the transforming factor had to be treated for protein removal again since the sugar-splitting enzyme which had been added is a protein. Alcohol was now added and a stringy mass which was free of protein and free of sugars separated out. When redissolved at a dilution of one part in 600 million the product was still potent in effecting transformation. The transforming principle was thus, at long last, tracked down and isolated in a form amenable to chemical studies. Tests on the solid itself confirmed the earlier conclusions. It was not a protein or a sugar. The very method of isolation ruled out the possibility of a fat, for no fat could remain after repeated extractions with chloroform and precipitation with alcohol. Of the known, major, solid organic components of a cell the only possible candidate left for the role of the transforming agent was one, or both, of the nucleic acids: RNA or DNA. The product, as it turned out, did contain phosphorus, a telltale indication that it might be a nucleic acid. A chemical reaction which discriminates between DNA and RNA was positive for the former. A complete analysis of the four elementary components of DNA, carbon, hydrogen, nitrogen, and phosphorus, was performed. (There was no convenient assay for the fifth large, elemental component, oxygen.) The report of the analyst, compared with the values calculated for DNA, is as follows:

	Carbon	Hydrogen	Nitrogen	Phosphorus
Found in Avery's Product	35.50	3.76	15.36	9.04
Theoretical for DNA	34.20	3.21	15.32	9.05

This and other analytical data conveyed an unequivocal message: The transforming principle was DNA.

Avery pondered the meaning of his finding:

In the present state of knowledge any interpretation of the mechanism involved in transformation must of necessity be purely theoretical. The biochemical events underlying the phenomenon suggest that the transforming principle interacts with the R[ough] cell giving rise to a coordinated series of enzymatic reactions that culminate in the synthesis of the Type III [smooth] capsular antigen. The experimental findings have clearly demonstrated that the induced alterations are not random changes but are predictable, always corresponding in type specificity to that of the encapsulated cells from which the transforming substance was isolated. Once transformation has occurred, the newly acquired characteristics are thereafter transmitted in series through innumerable transfers in artificial media without any further addition of the transforming agent. Moreover, from the transformed cells themselves, a substance of identical activity can again be recovered in amounts far in excess of that originally added to induce the change. It is evident, therefore, that not only is the capsular material reproduced in successive generations, but that the primary factor which controls the occurrence and specificity of capsular development is also reduplicated in the daughter cells. The induced changes are not temporary modifications but are permanent alterations.

That brief paragraph interprets the result of one of the greatest experiments in biology. The transforming factor, which is made of DNA, achieves two things in a bacterial cell: It becomes integrated into the genetic apparatus of the cell, and then it gives rise to a coordinated series of enzymatic reactions.

Avery did not state it, but in effect he had isolated the genetic material of the cell. Thus the two independent paths of research stemming from the discovery of DNA by Miescher and the deduction of the laws of heredity by Mendel converged. It is in DNA that the infinitely precious genetic information is inscribed. DNA is the blueprint of the cell.

Avery's stunning finding galvanized biological science. With the new information at hand a multitude of new questions both on the structure and the mode of action of DNA were posed. First of all, Avery's claims had to be confirmed. He was so meticulous an experimenter and defined his methodology so well that this presented no problem. Soon in many laboratories "transforming DNA" was being extracted from organisms other than the pneumococcus and recipient cells were being transformed to possess hereditarily a variety of attributes of the donor cells. For example, DNA can be extracted from a donor which is resistant to a certain drug and the drug resistance can be transferred to an otherwise sensitive culture. Or the capacity to form some pigment perpetually can be transferred to an erstwhile colorless organism. Scores of well-defined attributes in several different strains of microorganisms were shown to be under the seminal control of DNA. The conviction grew that in microorganisms, at least, the capacity to express inheritable traits is invested in DNA.

That DNA is the genetic material even in the tiniest specks of matter on the threshold of life, the bacteriophages, was soon demonstrated by Dr. A. D. Hershey of the Cold Spring Harbor Laboratories on Long Island. A bacteriophage has two structural components. It is made up of a protein bag and tail, and the bag is filled with DNA. The element phosphorus is an obligatory component of DNA but is absent from protein; on the other hand sulfur is absent from DNA but is present in many proteins, including the protein of the bacteriophage. What happens upon the addition of bacteriophage to a culture of bacteria can be visualized by a series of sequential photographs taken through an electron microscope. At such magnification a bacteriophage alongside a bacterium appears like a tadpole next to a walrus. Within a couple of minutes after the mixing of phage and bacterium the tadpole becomes attached by its tail to its large victim. (A bacterium surrounded by a phalanx of dead bacteriophage may be seen in Figure 1 Chapter 4.) About twenty minutes later the bacterium disintegrates and a hundred fresh bacteriophages emerge from the dead hulk.

With the aid of radioactive tracer methodology, the description of which is more appropriate in the next chapter, Hershey was able to show that only the phosphorus and not the sulfur of a bacteriophage penetrates during infection. Therefore, only the nucleic acid of the phage, not its protein, gets into a bacterium. Thus the bacteriophage's ability to capture the bacterial cell's machinery and to order it to make not its normal components but the substance and structure of the attacking phage must be invested in the DNA. Eight years after Avery's discovery, this was the first complete confirmation, with an entirely different system, that the master molecule is DNA not only in bacteria but in the phages as well.

Credit for this outstanding achievement of Hershey's should be shared, in part, by two other investigators on whose observations Hershey could build his ingenious experiment. Dr. Thomas F. Anderson of Philadelphia, an outstanding electron microscopist, had observed merely from the shadows cast by the bacteriophage that after they come in contact with bacteria they become flat, empty bags. A similar disgorging of their contents can be achieved by merely shaking the bacteriophage in distilled water. But in this case the disgorged material is amenable to analysis. Dr. Roger Herriot of Johns Hopkins University had shown that under such conditions DNA oozes out of the bacteriophage.

Hershey then performed his beautifully conceived and impeccably executed experiment.

How about the viruses which contain no DNA at all, such as the plant viruses which contain only protein and RNA? Where is the information for their replication stored? Soon after Hershey's experiment the agent of infectivity in an RNA virus was also revealed. This was done with tobacco mosaic virus, which has occupied a stellar role in virus research from the very beginning. Indeed, it was the first virus to be discovered. The distinction between a bacterium and a virus was made by the Russian bacteriologist Dmitri Iwanowsky in 1890. He traveled to the Crimea, where the tobacco fields were infested with a disease that mottled and withered the green leaves. The Tatar tobacco growers of the region called it the marble disease. In the eyes of more worldly observers the mottling recalled a mosaic pattern and that image gave the widely accepted name to this curse of tobacco growers: the tobacco mosaic disease. Iwanowsky searched for the infective agent by filtering the juice of a macerated diseased leaf through an unglazed porcelain cup. This device served as an impenetrable sieve for all known bacteria. But in this case it permitted the infectious agent to pass through it: in other words the clear filtrate was just as infectious when painted on tobacco leaves as was the original unfiltered mash. The term filterable virus was given to this elusive agent of infection. Soon the hoof and mouth disease of cattle was also shown to be caused by a filterable virus, and after this a whole host of diseases was shown to have such elusive etiological agents. When we are invaded by the appropriate virus we may come down with rabies, smallpox, yellow fever, dengue fever, infantile paralysis, measles, mumps, influenza, virus pneumonia, and the common cold. The bacteriophage which was discussed above in connection with Hershey's work also belongs in this category.

In 1935 Dr. Wendell M. Stanley, a young organic chemist working at the Rockefeller Institute, decided to investigate the chemical nature of the viruses. Only a few years earlier several enzymes were purified sufficiently to yield homogeneous, pure crystals. In such a pure state these enzymes could be meaningfully analyzed, and they all turned out to be proteins. Stanley extracted tobacco leaves mottled with the mosaic virus and subjected the juice to the same chemical manipulation which had yielded crystalline proteins. The intellectual climate of the time conditioned Stanley to look for a protein as the active agent. Indeed, he isolated what he thought was a protein no different from other crystalline proteins in appearance. If a water solution of this material even at a dilution of one part per billion was applied to a healthy tobacco leaf, in due time the broad green leaf became a mottled, bedraggled shadow of its former self. A British biochemist, Pirie, demonstrated large amounts of phosphorus in Stanley's preparation and this was a sure sign of the presence of a nucleic acid. Actually the virus turned out to be a miniscule amount of RNA wrapped in an abundant protein coat.

An example of the preoccupation with proteins as the active agents in viral infection is given by Dr. Heinz Fraenkel-Conrat, a distinguished investigator in this area. In 1952, when he joined the Virus Laboratory in the University of California at Berkeley, he initiated experiments in which he tried to produce viral mutations by altering the protein of the tobacco mosaic virus by chemical means. However, he quickly recovered from these initial fumblings and became the first to show that the infectivity of this virus is within the structure of the RNA. Essentially what Fraenkel-Conrat achieved was a chemical dissection of the tobacco mosaic virus into its protein and nucleic acid components. But he also learned to reassemble them into infectious particles. Tobacco mosaic virus has several mutants which can be recognized by the individual mottling pattern they produce on the tobacco leaf. If in the reconstitution experiments hybrids were formed by taking the protein from one type of virus and the

RNA from another, it was always the characteristic attribute of the virus from which the RNA was derived that was expressed by the man-made hybrid. This was a clear indication that the inheritable characteristics of the mosaic virus were entrusted somehow to the RNA. Even more conclusive evidence was soon provided by Fraenkel-Conrat and by Dr. Schramm of Tübingen, Germany. They both succeeded independently in preparing pure RNA, devoid of protein, which was still infectious when painted on a tobacco leaf.

Three different lines of evidence now pinpointed nucleic acids as the genetic material. Transforming DNA, bacteriophage DNA, and tobacco mosaic virus RNA are each the active agent in transmitting hereditary information into the target cells they invade.

But how do these nucleic acids achieve their purpose? How do they subvert the host cell and force it to fashion not its own normal components but the stuff and structure of the virus? An unequivocal answer to this question came from Dr. Seymour Cohen, a biochemist at the University of Pennsylvania. First of all, Dr. Cohen and his associate, Dr. Wyatt, discovered a fundamental difference in the DNA of certain bacteriophages and in the host bacteria they invade. In the DNA of bacteria one of the four cardinal components is cytosine.

Cytosine

This base is absent from the DNA of certain bacteriophages, one of which has been arbitrarily designated as T_2. Instead of cytosine the DNA of T_2 bacteriophage contains hydroxymethyl cytosine.

Hydroxymethyl cytosine

Both the substance, hydroxymethyl cytosine, and the apparatus for its assembly are lacking from normal bacteria. However, Dr. Cohen found that very soon after invasion by the phage a new enzyme which is capable of performing this task begins to accumulate in the subjugated bacterial cell. Thus the phages—whose name is derived from the Greek word to eat—can not only "eat" bacteria, but they can order the installation of apparatus for new food to be cooked.

Dr. Cohen thus provided one more evidence, if that were needed, that genetic materials achieve their seminal tasks by the assembly of new protein molecules.

But how does one component of the cell, a nucleic acid, make another one, a protein, which is so completely different both in structure and function? I will ask the reader to be patient. Read one more chapter on DNA structure and then we shall explore protein synthesis.

Dr. Avery's discovery, which might be symbolized as

$$DNA = \text{Hereditary Information,}$$

was to have an impact on biology equal to that of Einstein's

$$E = mC^2$$

on physics. Yet he died in 1955 in almost complete obscurity. The Nobel Prize Committee has so far honored three men for deducing the structure of DNA, another one for demonstrating its mode of synthesis—but Avery, whose contribution towers above all this, was never reached.

But this is the price geniuses often pay for being too far ahead of their contemporaries; by the time their work is appreciated they may be gone. Yet they are not to be pitied, for with the ability to see what others miss often comes an inner vision, too, of their own worth. Avery *knew* the value of his discovery and he therefore passed the criteria of excellence for his most severe judge —himself. But more than this, he had the ultimate reward of a scientist. He saw in an exultant moment of transcendent brilliance what had lain deeply hidden through the eons of time since life began. He saw the blueprint of life.

7 THE ARCHITECTURE OF NUCLEIC ACIDS

II. THE SPIRAL OF LIFE

CREATURES OF CIRCUMSTANCE is what Somerset Maugham calls
one of his collection of short stories. Of course his title refers to
the denizens of a limited tropical world who are shaped by the
triple circumstances: the tropics, boredom, and alcohol. But
while his heroes—if such they be—are limited, his description of
mankind—and indeed of all things living—is universal: we are
all creatures of circumstance. We have been shaped by the
confluence of chance forces which happen to prevail on this, our
mother planet, in whose vast seas all things living had their
origin.

Consider the very structure of water itself. That water is
important to life is obvious. It covers three fourths of the surface
of the planet and it composes 65 percent of our body. But it is
only the rarest chance which permits water to be the life-giving
and life-shaping liquid that it is. If not for this chance attribute
water would be a gas at the temperatures now prevailing and our
planet would be a gas-enshrouded, hot, arid, absolute desert
devoid of any moisture and, consequently, of any life.

What is this attribute, then, to which we owe everything, the sweet climate of our planet and our very life? It is a very simple chemical propensity: the ability of hydrogen and oxygen atoms to associate with each other in excess measure. It is a general rule that elements belonging to the same family in Mendeleyev's periodic table behave quite similarly in chemical reactions. Thus oxygen and sulfur, which are eight elements apart from each other, both form a dihydride. The dihydride of oxygen is, of course, water—the dihydride of sulfur is H_2S, hydrogen sulfide, the malodorous gas which issues from rotten eggs or from some sulfur springs. Now, it is also a general rule that the heavier compound of such a homologous pair invariably has the higher boiling and freezing points. And here is where the unique attribute of water enters. H_2S has a molecular weight of 34, while water has one of 18, or the former has almost twice the mass of the latter. Yet the boiling point of water is 161°C above that of H_2S. Had water behaved as could be anticipated from the attributes of the other dihydrides of this family of elements its boiling point would be not 100° Centigrade but −80°. And what a different world this would be as a consequence! In the first place the planet would not have cooled to its present temperature. It has been the repeated cycles over the millenia, of cascading rains and instant evaporation of the total contents of all the oceans, which have brought our planet from a molten ball of lava to its present ambient temperature. Moreover, had water behaved as H_2S does, its vapors could not have condensed to form the seven seas until the average temperature had reached −80°. (This is colder than the inside of the most efficient of home deep freezers.)

It would have taken millions if not billions of years hence to reach such a low temperature, and once reached it would have been an inhospitable environment for the cumulative chemical reactions which eventually might yield that miracle of molecular evolution, a living cell.

To what do we owe the mild climate of our planet and the pleasant solidity of our bodies, 65 percent of which might have been a gas? We owe it all to the tendency of hydrogen and oxygen to form an unusual bond, the so-called hydrogen bond. Hydrogen sulfide (H_2S) exists as a discrete independent molecule and, therefore, it remains a gas until it is cooled to $-61°C$. Water, on the other hand, becomes associated with other water molecules by the formation of extra bonds between the oxygen of one molecule and the hydrogen from a neighboring water molecule, thus:

In this way, three molecules of water become condensed into one, forming a much more sluggish molecule, which boils at $100°$ instead of $-80°$. Such a substance had the capacity to prepare our earth for conditions which were propitious for the molecular evolution which led to life.

This, of course, is the most pervasive influence of hydrogen bonding, but as we shall soon see hydrogen bonding plays a dominant role in the shaping of nucleic acid structures as well.

But before we can take up the impact of hydrogen bonding in this area we must pick up the thread of the development of nucleic acid structure, which we left in Chapter 3.

As we saw in the previous chapter, two stunning biological discoveries focused attention on the neglected area of nucleic acids. The first of these occurred in 1935 when Dr. Wendell M. Stanley crystallized the tobacco mosaic virus. The second was Avery's discovery, nine years later, which implicated DNA as the carrier of genetic information.

Biochemists now began to flock to the study of nucleic acid chemistry and they brought badly needed newly developed tools

with them. The four nucleotides which compose nucleic acids
are very similar in chemical and physical attributes, rendering
their separation so tedious and uncertain that Levene's tetranucleotide hypothesis, which claimed the presence of the four bases
in DNA in equivalent amounts, went unchallenged for years for
the lack of better data. (Actually, as we shall see, there are wide
variations among different species of organisms in their base
ratios.)

Newly developed analytical tools soon undermined Levene's
hypothesis. The first to seize on paper chromatography for the
analysis of nucleic acids were Dr. Erwin Chargaff of Columbia
University and a Swiss postdoctoral visitor in his laboratory,
Dr. Ernst Vischer. With this simple technique they opened up
vast new frontiers of research in the origin, structure, and
function of nucleic acids. In the first place the tetranucleotide
hypothesis was buried, for even among such closely related
organisms as bacilli—which belong to the same genus—there
may be as much as 100 percent variation in the content of a
specific base in their DNA. But from Dr. Chargaff's painstaking
analyses an entirely different—and real—pattern emerged. In
all the DNA's tested, the number of molecules of guanine always
equals those of the cytosine, and the adenine always balances the
thymine.

Guanine   *Cytosine*

Adenine   Thymine

The meaning of the quantitative equality of the

$$O \quad \text{and} \quad NH_2$$
$$\| \qquad\qquad |$$
$$C \qquad\qquad C$$

groups (on the top of the four structures) eluded its discoverer but was soon comprehended by two investigators in a flash of brilliant insight.

Another tool brought to the study of nucleic acid structure is a very clever variation on Tswett's column chromatography, which was discussed in Chapter 5. Dr. Waldo Cohn is a chemist who worked on the Manhattan Project, which developed the atom bomb. He studied separation of elements by what has come to be known as ion-exchange chromatography. This technique makes use essentially of the Tswett column, but instead of a relatively inert substance like limestone a highly charged granular material—usually some negatively or positively charged artificial plastic—is used to pack the column. The mixture to be separated is absorbed on this column and is then eluted differentially not by an inert solvent but by increasingly acidic—or basic—fluids. Hence the name "ion-exchange" chromatography—the ions which are absorbed onto the column are exchanged for the more strongly charged ions of the solution.

When the Manhattan Project achieved its goal, Dr. Cohn and his method suffered technological unemployment. But not for long. Being a very versatile gentleman—he is almost as good a conductor of symphony orchestras as he is a chemist—Dr. Cohn decided to apply his tool to nucleic acid chemistry. That he was permitted to do this is a monument to the wisdom of his superiors at the Oak Ridge Laboratories, for Dr. Cohn's switch of interest brought rich bounties in our understanding of nucleic acid structure. The most fundamental contribution Dr. Cohn made was the elucidation of how nucleotides are strung together to build the giant chains which hold in their sequence the precious information of shaping a living cell.

When nucleic acids are dismembered under certain conditions nucleotides are obtained. These contain a base, a sugar, and phosphoric acid. An example of such a nucleotide might be uridylic acid obtained from RNA:

This would be called the 5′ phosphate because the phosphoric acid is attached to carbon 5 of the ribose.

But with Cohn's delicate analytical methods it could be shown that under some conditions uridylic acid can exist not only as a 5′ phosphate but as a 3′ phosphate in which the phosphate is attached to carbon 3. This is shown on the next page.

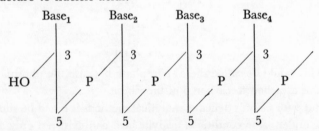

This finding revealed the method of accretion of nucleotides into the giant nucleic acid molecules. A bridge is formed via a phosphoric acid between carbon 3 and carbon 5 of two adjacent sugars. Therefore, since there is only one phosphoric acid between two sugars, it can go to either of the two—to the 3 or 5 position—upon cleavage, and Cohn's method of chromatography can discern the extremely subtle difference between the two.

This is the basis for our assigning the following schematic structure to nucleic acids.

This is as far as the conventional tools of organic and of analytical chemistry could take us in the development of our ideas on the structure of nucleic acids. While the determination of these structures was a great achievement, they contributed nothing to our understanding of the biological functioning of nucleic acids.

How is genetic information encoded in DNA? How is DNA replicated as a cell divides so that identical genetic blueprints can be dowered out to both daughter cells? The answer to these questions which are uppermost in the mind of a biologist must reside in the structure of DNA. But there are more subtle questions as well: How are the genes—and therefore the DNA—divided in the formation of the sex cells? And once sexual fusion has taken place how does the gene from one parent overpower its counterpart from the other sex cell to achieve dominance in the expression of a trait?

The classical biologists rarely attempted to correlate biological function with the structure of the cell's molecular components. For example, T. H. Morgan said in 1934 in his Nobel Prize lecture:

There is no consensus among geneticists as to what genes are— whether they are real or purely fictitious—because at the level at which genetic experiments lie it does not make the slightest difference whether the gene is a hypothetical unit or whether the gene is a physical particle.

But the biochemist cannot accept disembodied hypothetical units. He must strive to isolate such a factor and decipher its chemical structure, for it is an article of faith with him that the secrets of the mechanism of life are locked within the physico-chemical structures of the molecules that are the edifice of life. With respect to DNA the problem placed before us is the deciphering of a chemical structure which can explain—or is at least compatible with—its known biological functions.

For this task still newer and more subtly probing tools had to be applied to DNA. As we pointed out in Chapter 2, a whole school of X-ray crystallography was founded in England by the gifted father and son W. H. and W. L. Bragg. One student of this school, William Astbury, had the courage to tackle the forbidding problem of the X-ray crystallography of the macromolecules of the cells, the proteins and nucleic acids. As early as 1938 Astbury came to the conclusion that DNA has a linear, repeating structure in which flat planes are inserted at right angles to the long axis of the molecule.

In 1951, as was pointed out in Chapter 5, Linus Pauling deduced the helical structure of proteins and two years later extended his conclusions to the structure of nucleic acids. Pauling and his collaborator, Corey, proposed a helical structure for DNA—something like a spiral staircase. According to this model the phosphoric acid groups form the core and the bases radiate outward as do the steps of a spiral staircase. This model failed to incorporate a very suggestive bit of information which had been provided by the British biochemist J. M. Gulland. He came to the astute conclusion that the bases of DNA must be extensively hydrogen bonded to each other. Three new investigators now took up the profoundly challenging jigsaw puzzle of the structure of DNA. (Since the Nobel Prize Committee lumped them together so can we.) They are M. H. F. Wilkins, F. H. C. Crick, and J. D. Watson. The first two are English physicists who worked on the atom bomb project and on the design of aerial torpedoes and who at the end of the war found themselves the victims—or rather, from their point of view the beneficiaries —of technological unemployment. A great many physicists chafe under conditions where their work is part of a group effort. If given a chance they run to biology with the eagerness of schoolboys running to a playing field after the day's restriction in the classroom, for in biology it is still possible to work alone. Wilkins and Crick searched and found a new area to engage their talents: the structure of biological macromolecules. Dr. James D. Watson is an American biologist who became frustrated with the limitations of the tools and ideas of classical biology and went to Cambridge to learn X-ray crystallography. Watson and Crick found each other intellectually congenial and began to poke around the problem of the structure of DNA.

The ease with which scientists join into working partnerships after the briefest of personal acquaintanceship comes as a surprise to those not involved in science. But this is a natural consequence of the nature of the profession: ours is a lonely pursuit. We live alone in a world of phantasy. No one else but those with training and imagination close to ours can become a friend—or a foe—in this phantasy world. I prize a close friendship of over thirty years' duration with a distinguished lawyer. Whenever we meet he will ask in the middle of the first martini: "How's your work?" My answer: "So, so" or "O.K." disposes of that topic for the rest of the evening. But I can meet for the first time a Bulgarian colleague and despite the difficulty of expressing complicated thoughts in primitive, basic English, within minutes we are immersed in the intimate world of our ideas.

Three factors have contributed immensely to the growth of the physical sciences. (1) The universal acceptance of clearly defined terminology. (A 1.0 molar HCl solution is identical whether made up in Minneapolis or in Moscow, but try to get an agreement between two political scientists from those two locales on the meaning of democracy.) (2) The tradition of publication of new discoveries with the obligation to furnish sufficient details to enable anyone skilled in the area to repeat it—if necessary with the original author's materials. (We often literally cut our own throats by sending to prospective competitors our unique organisms or compounds. The only solace is the conviction many of us hold that the advancement of science is more important than our own.) (3) The easy camaraderie among scientists the world over, which is rooted in the scarcity of those with whom we can really discuss the largest part of our life: our work.

Watson and Crick had the following bits of the jigsaw puzzle to work with. Wilkins had repeated with the latest refinements the early work of Astbury on the X-ray crystallography of

DNA. From the patterns of the dispersion of the X rays he came to the following conclusions. The DNA molecule is made up of regularly repeating units, and the molecule is helical. The regularity of the repeating structures presented a paradox. Chemical analyses indicated that the purine and pyrimidine bases come in random sequence, yet from this randomness a repeating regularity emerged in the X-ray pictures. Wilkins made this information available to his colleagues at Cambridge.

Watson and Crick were aware of the analytical findings of Chargaff which indicated the quantitative equivalence of guanine and cytosine and of adenine and thymine. Finally, Gulland had emphasized the presence of extensive hydrogen bonding and had even suggested that the chains of DNA might be linked together by hydrogen bonds to form multichain complexes.

Watson and Crick now brought a still newer tool to the task: conformational analysis. This intellectual pastime is an outgrowth of the information about the structure of atoms and molecules which the X-ray crystallographers had gathered. The actual size of the various atoms is quite accurately known and the distances between atoms when they are present in molecules have been carefully measured. From this information we can build accurate scale models of atoms with little connecting tie rods that reproduce the appropriate bond distances. Scientific research has been facetiously called organized play for adults; molecular kits are our Erector Sets. With such atomic models the plausibility of various molecular structures can be explored. For if atoms cannot reach each other or are too bulky for a given site in the scale model their existence in a stable molecule is also highly unlikely.

Watson and Crick made scale models of the four bases of DNA and they found that if they brought together (as if bonded together by hydrogen bonding) adenine and thymine and guanine and cytosine the over-all dimensions of these two couplets were identical. This was cause for rejoicing, for according to Chargaff's data the units in each of these couplets are present in equivalent amounts and if they formed such pairs in DNA such analytical equivalence would be obligatory.

The formation of such hydrogen-bonded pairs with identical over-all dimension would resolve a paradox presented by the X-ray data: the emergence of nearly identical repetitive structures out of the disorder of the random sequence of bases; the repeated units are not single bases of different sizes but base pairs of identical dimensions. The hydrogen-bonded structures of the two pairs of complementary bases are shown below.

The Watson-Crick model of DNA—as it is called—is, then, a double helix with each chain twisted around the other. The sequence of bases in each chain complements that of the other. Below is a two-dimensional representation of a segment of DNA.

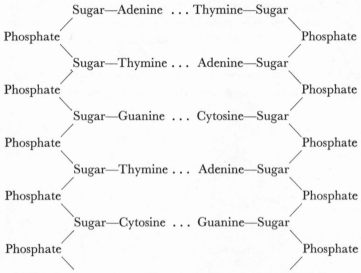

Complementary sequences of bases in the 2 strands of the double helix.

Figure 7.1 is a photograph of the first Watson-Crick model, which the progenitors of it had fashioned out of crude wire. Figure 7.2 is a photograph of an atomic model and its schematic representation.

Figure 7.3 represents still another atomic model plus its representation with the aid of the symbols of the four suits of a deck of playing cards. This model was exhibited at the Seattle World's Fair. In ten years the model of DNA has emerged from the ivory tower of the laboratory into the hurly-burly of public exhibitions.

This brilliant hypothesis is consistent with all the known physical attributes of DNA and in addition it makes a prediction about an important biological function of DNA: the mode of replication of DNA itself.

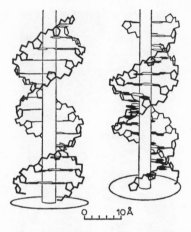

Fig. 7.1. *Diagrammatic drawing of the Watson-Crick helix of DNA. The lines represent the phosphate ester chain and the disks between them indicate the paired bases.*

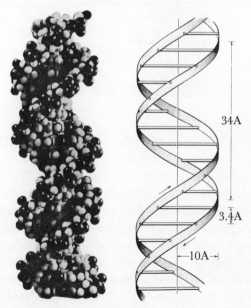

Fig. 7.2.

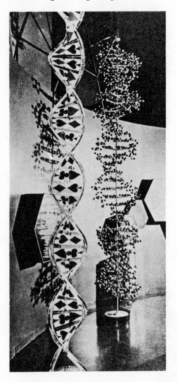

Fig. 7.3.

Prior to cell division the DNA content of a cell must be doubled so that in each daughter cell faultless copies of the total genetic material can be invested. A structure of DNA with two identical helices fused by hydrogen bonding provides an ideal model for such replication. If the two strands separate as the two halves of a zipper and each twin coil serves as a model on which a complementary coil can be fashioned—i.e., a thymine vis-à-vis an adenine, and guanine vis-à-vis a cytosine—then the exact replication of the DNA and, therefore, of a gene is ensured.

Such a method of replication of DNA is represented schematically in Figure 7.4. The capital letters represent the bases (T for thymine, A for adenine, C for cytosine, and G for guanine; P is for the phosphate bond and the pentagons for deoxyribose).

This concept of the replication of DNA was revolutionary, for it assumed that the DNA acts as a template and a directive force on some enzyme, inducing it to manufacture more of the template. Such an enzyme was unknown in our total biochemical experience. To demonstrate its existence required a man of unusual intellectual power to conceive it, of enormous self-confidence to undertake it, and complete self-discipline to achieve it. Dr. Arthur Kornberg—now of Stanford University—was the man who was equal to this task.

The isotopic tracer technique was a *sine qua non* for Kornberg's success. Let us review briefly the history of this powerful tool.

Isotopes of an element contain the same number of protons

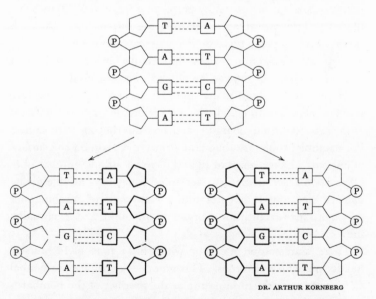

DR. ARTHUR KORNBERG

Fig. 7.4. Proposed scheme of replication of a Watson-Crick DNA model. Boldlined polynucleotide chains of the two daughter molecules represent newly synthesized strands.

and, hence, planetary electrons, but differ in their endowment of neutrons. They thus have the same atomic number and essentially similar chemical properties, but their atomic masses differ. For example, the vast majority of hydrogen atoms have an approximate mass of 1 on the atomic weight scale. But there is one atom of hydrogen in about 5000 that weighs twice as much. In the case of nitrogen most of the atoms have a mass of 14, but one atom in 270 has a mass of 15. Methods are available for the concentration of the less abundant isotopes and for their accurate measurement. This is done in the mass spectrometer, an instrument in which elements are converted into ions which are then drawn into the field of a powerful magnet. In the magnetic field the charged particles are deflected from their path in proportion to their mass. The degree of deflection—and, therefore, the relative mass—is measured by permitting the ions to fall into slits placed at intervals along the path of the surging ions. At each slit a charged plate receives the ions cascading on it and by appropriate electronic devices the relative amount of ions of different weights in the various slits can be measured.

There are a great many isotopic species whose structure renders them unstable and which therefore decompose and emit in the process various charged particles, thus they are radioactive. For example, the most abundant atoms of carbon on the surface of our earth have a mass of 12 and have an atomic number of 6. Of carbon's six planetary electrons, four are the life-giving valence electrons that enter into the myriads of combinations which produce a multitude of organic compounds and, eventually, life.

There is an isotope of carbon which has a mass of 14 but still has six planetary electrons. This isotope—for which the symbol C^{14} is used—is present in nature as the product of the bombardment of our atmospheric nitrogen (N^{14}) by cosmic rays. It can also be made with the vast energy of an atomic reactor. Carbon 14 has an unstable structure and keeps decomposing steadily, emitting a constant stream of β particles. These can be very conveniently measured with a Geiger counter, which is designed to be sensitive to such disintegrations.

It is thus easy to tell apart two isotopes—both the stable and unstable—by physical means. But can the fastidious enzymes in a living cell discern the difference between isotopes? If they do then one or the other of two species of isotopes of a given element should be more concentrated in the cell than in the inorganic world. We can examine the nitrogen atoms bound in the proteins of our hair. Prior to their incorporation into our body those nitrogen atoms may have sojourned in the cells of countless plants and animals, including humans.[1]

During their cycle through plants and animals the nitrogen atoms may have participated in countless enzyme-catalyzed reactions, and if the enzymes could discriminate between isotopes one of them should have been concentrated. But the proportion of N^{15} in the nitrogen of our hair is exactly the same as that of the nitrogen in our atmosphere. This is the fundamental basis for our ability to use isotopic tracers in biological studies.

The pioneer in the use of isotopes for biochemical research is Dr. David Rittenberg of Columbia University. The methodology developed by Rittenberg and his colleague, the late Dr. Rudolf Schoenheimer, is the foundation on which is based all current biological tracer work—and this has grown to vast proportions. The use of isotopic tracers has contributed more than any other tool to the revolution in biological knowledge which has taken place during the past thirty years. Biochemistry was a static descriptive science prior to 1935. Scores of the molecular components of the cell were isolated and identified, but of their interactions within the living organism next to nothing was

[1] Why hasn't anyone written a historical novel with an atom as its indestructible, ubiquitous hero? Imagine what a nitrogen atom which had been a corporal part successively of Cleopatra, Napoleon, and Franz Liszt—to chose randomly three flamboyant characters—could put in its memoirs! The title inevitably would be "Forever Atom."

known. The biosynthetic pathways which shape the numerous components of the cell were almost completely obscure. We could isolate cholesterol, measure its level in the blood, but its origin from acetic acid, a product of carbohydrate metabolism, could not even be guessed at. Without isotopic tracer techniques a study of the metabolism of the body's components, which are continually synthesized and broken down, presented insurmountable difficulties.

Questions which formerly could not even be formulated are answered routinely with the aid of isotopes. The origin of the building units of the complex molecules of a cell and the exquisitely intricate methods of their assembly have been revealed with the suddenness and splendor of a sunrise over a hidden, dark landscape.

With the aid of tracer methodology the minute amount of synthesis of complex molecules by cells removed from integrated tissues, and, indeed, by mere extracts from such cells, could be demonstrated. By the time Dr. Kornberg started his attempts to synthesize DNA there were successful achievements of the partial synthesis even of proteins to encourage him. The chronology of these milestones, written by one who attempts to be an impartial historian, is as follows. In 1942 Dr. Rittenberg and Dr. Konrad Bloch—now of Harvard University—found that the relatively simple compound acetic acid, which contains but two carbon atoms, is the building unit out of which animals fashion the complex compound cholesterol, which contains twenty-seven carbon atoms. Two years later the same workers, with assistance from the writer of these lines, showed that the synthesis of cholesterol could be achieved by cells isolated from liver. Prior to this, only degradative reactions could be demonstrated by isolated mammalian cells, and thus this was the first achievement of the synthesis of a complex structural component of the isolated cells.[2]

---

[2] Since these lines were written Dr. Bloch received the Nobel Prize in Medicine in 1964 for this and subsequent work.

Dr. Nancy Bucher of Harvard University extended our findings by a bold leap. She found that cholesterol can be fashioned from isotopically marked acetic acid even by disintegrated cells, provided a source of stored energy is supplied. The lady's achievement paved the way for more complex undertakings. For example, Dr. Paul Zamecnik, also of Harvard University, has paid tribute to Dr. Bucher's finding as a source of encouragement to him in undertaking attempts at protein synthesis in cell-free extracts. As we shall see in the next chapter Dr. Zamecnik and his collaborators have been the unquestioned leaders in this area. By 1953, when the Watson-Crick hypothesis was proposed, they had achieved protein synthesis with cell-free preparations. The creation of DNA with enzyme extracts, was reported three years later by Dr. Kornberg. I chose the word creation for this achievement advisedly, for it is the belief of many, among them H. J. Muller, Nobel laureate biologist, that with the synthesis of DNA Dr. Kornberg accomplished nothing less than the creation of life.

Dr. Kornberg had extensive experience with the nucleotides which serve not only as building units of nucleic acids but also assist in many enzyme-catalyzed reactions as so-called coenzymes. One of these nucleotides, adenosine triphosphate (ATP), is the ubiquitous source of energy for many of the chemical reactions in the cell which require energy to drive them to completion.

Most of these nucleotides have a phosphate group in the 5 position of ribose, and enzymes are present in all cells which can fashion the 5 phosphates of deoxyribose-containing units as well. From this information and from his experience Kornberg came to the intuitive conclusion that the starting materials for the assembly of DNA by enzymes must be the four bases attached to deoxyribose in whose position (5) a triple phosphate group is inserted.

Such triphosphates contain large amounts of chemical energy

and, therefore, no other external source of energy would be required to carry the reaction forward toward the assembly of DNA.

An extract from a cell contains enzymes that decompose nucleic acids, so Kornberg could expect no net increase in the total DNA from his early attempts. He therefore resorted to the use of carbon-14-labeled nucleotides. Twenty years earlier, the preparation and patient assembly of such products would have taken months, but fortunately by the mid-1950s such compounds were commercially available. This was one of the bounties which had accrued mainly from the financial support which our federal government started to channel toward research at the end of World War II. In the area of biochemistry such funds were meted out with extraordinary efficiency and wisdom largely by the National Institutes of Health. The superlatives represent not only the opinion of the writer. For example, Dr. Fritz Lipmann of the Rockefeller Institute, Nobel laureate in biochemistry, speaks of the grant system of the National Institutes of Health as an "almost miraculously successful government expenditure."

With the funds from these sources the researcher was able to purchase research chemicals which theretofore were both beyond his means and in most cases unavailable. But a large, supporting industry grew up around biochemical research and today the rarest of research compounds, with isotopic labels in appropriate positions in the molecule, are stocked and sold by several thriving supply houses.

Kornberg added some DNA and four triphosphate precursors —one of them bearing radioactivity—to an extract prepared from the bacteria *E. coli*. After incubation, the DNA was separated from the solution by the addition of acid that renders it insoluble. The insoluble DNA was thoroughly washed to remove any radioactivity of the starting material itself and the scrubbed DNA was placed into the counting chamber of a Geiger counter. Out of an initial input of a million counts the DNA had acquired about fifty. A result like that is a supreme challenge to the researcher's strength. Most of us would have considered that experiment a total failure and prudently looked for a more productive area. But not Kornberg. Although disappointed by the miniscule yield, he thought it was real. As he put it: "Through this crack we tried to drive a wedge, and the hammer was enzyme purification." Today his enzyme preparations are several thousandfold more pure and DNA can be made with them in yields sufficiently large to dispense with the tracer technique for the proof of synthesis: The newly formed DNA accumulates in sufficient amounts to isolate it and study it chemically, physically, and eventually biologically.

The most exhaustive studies of the enzyme-made DNA confirm the suggestion implicit in the Watson-Crick model that the replication of DNA proceeds by the alignment of bases complementary to those present in the template DNA, as shown on page 169.

The ultimate triumph has recently been achieved. A DNA with biological activity has been synthesized. A DNA—containing bacterial virus, called phage Ox, has been synthesized in Dr. Kornberg's laboratory with the aid of the enzyme he discovered. We can now agree with Dr. Muller that life has been created by man. To be sure, we must carefully circumscribe our definition of life. We must not expect the creation of a whale. But if we define life as an entity which is capable of producing order out of disorder at the expense of energy, and possesses the capacity for mutability, then DNA is alive.

The Watson-Crick hypothesis leaves many questions unanswered. Do both strands carry information? What is the mechanism of the reduction of the number of chromosomes to one half during the formation of the sex cells? How does one gene gain ascendancy over another and express itself as the dominant one? The hypothesis offers no illumination of these questions.

But it is too much to expect any hypothesis—no matter how brilliant and how penetrating—to explain everything completely. We must remember that Newton's hypothesis on gravitational force was proposed 300 years ago and it has withstood every challenge and gained confirmation in abundance—and yet today we have not the faintest idea *what* gravitational force is and from where within the structure of matter it originates.

"A mighty maze but not without a plan."

ALEXANDER POPE

8 THE WEAVING OF A PROTEIN

THE FATUOUS riddle, "Which came first the chicken or the egg?" has a counterpart at the molecular level: "Which came first nucleic acids or proteins?" The analogy between the two is complete, for each riddle probes the origin of the components of a reciprocal, seminal relationship. The function of nucleic acids is to give rise to proteins; in turn, the function of some of those proteins is to give rise to more nucleic acids. We do not know the primeval order of the creation of these components of the machinery of life—nor shall we, unless a visit to Mars or some other planet may yield such a revelation—but we have learned a great deal about the almost miraculous interplay of nucleic acids and proteins in a contemporary living cell.

The first suggestion of the possible involvement of a nucleic acid in the shaping of protein molecules came around 1940 from the Belgian embryologist J. Brachet and the Swedish biologist Caspersson. It has been observed by histologists that rapidly growing tissue had great affinity for basic dyes and therefore such tissue must contain acidic components. Caspersson had designed a special microscope in which the optical components were made out of quartz rather than glass. The device was well worth its cost, for with such a microscope living cells can be observed not only in the visible range of light waves, but in the ultraviolet as well (glass is opaque to ultraviolet light.) Nucleic acids are the only large components of the cell which absorb ultraviolet light intensively.

Therefore, under a microscope with quartz optical parts nucleic acid particles become pinpointed as opaque areas. Caspersson could thus identify the acidic components in rapidly growing cells as granules of nucleic acid.

Brachet made his correlation of nucleic acid levels and protein synthesis differently. For example, he observed that cells which make proteins for export are very rich in RNA. The silk gland of the silkworm, whose single-minded function is the production of a protein, silk, is the organ that is richest in RNA. On the other hand, the heart, an organ whose cells will make proteins only for repair, is one of the tissues with the lowest nucleic acid content.

The work of Caspersson and Brachet should have focused our attention on the involvement of nucleic acids—at least of RNA—in protein synthesis. But other ideas dominated the field at the time. In superficial hindsight it always appears that science moves from point to advanced point in a direct line to ever-increasing understanding of the world around us. This is a fallacy which must be set right. We owe this as a duty to history, and also because the contemplation of the errors of the past might make us more critical of the contemporary dogmas of science and more receptive to the ideas of those who will not ride the crest of the current vogue. For it must be remembered that in science as in any other human endeavor the participants are all too human. There are those who are ready to jump on a rolling bandwagon and bask in the security of working on a popular hypothesis approved by the pillars of the profession. But there are those who, endowed either with real imagination or just with a perverse streak, will refrain from joining the crowd and will endeavor to hack out a path of their own.

A prominent hypothesis of protein synthesis held about a quarter of a century ago is so sterile in retrospect that it is almost embarrassing to relate it. It was proposed by Max Bergmann, a great organic chemist at the Rockefeller Institute who unfortunately had little contact with biology. According to his hypothesis, proteins were not made of highly individual permutations of amino acids; the amino acids were supposed to be arranged in simple repetitive ratios. Sanger's brilliant studies of amino acid sequence in proteins soon laid this hypothesis to its well-deserved rest.

The postulated mechanism of protein synthesis had even less contact with biological reality. The enzymes which degrade proteins were supposed, under appropriate conditions, to reverse themselves and achieve the synthesis of new proteins. This primitive suggestion entrusted the shaping of the biological individuality of proteins to the reversal of the action of a series of degradative enzymes.

It is just as if a wrecking crew whose sole tool is one of those mammoth steel balls swinging from a crane were asked after they smashed an elaborate edifice to smithereens to rebuild it, using the same tool and without a blueprint of the original structure.

This scheme excluded any possible role for the nucleic acids in protein synthesis—whereas today we know this is their *prime* role. Such conclusions derived from work with highly purified, artificial systems could not possibly reveal the exquisite complexity of protein synthesis.

Actually, the Bergmann hypothesis marked the end of an era in biochemistry. The beginnings of that era reach back about seventy-five years to Sir Frederick Gowland Hopkins, who might be considered the founder of modern biochemistry. When Sir Frederick decided to study the chemistry of living things he was warned to desist: "The chemistry of the living?" said a timid colleague . "That is the chemistry of protoplasm; that is super-chemistry; seek my young friend for other ambitions."

Fortunately, Hopkins was not easily dissuaded; he started work on the chemistry of living things. But he was a pragmatic opportunist; he purposely focused on the small molecules in the realm of biochemistry. His strategy proved fruitful. He was the first to show the existence of food factors other than fats, proteins, and carbohydrates. He discovered vitamins. The generations of biochemists who have followed in his footsteps have isolated and identified hundreds of other chemical components from the living cell. These efforts in descriptive biochemistry have yielded rich rewards. Applying the knowledge patiently gathered by the biochemist we have all but conquered the deficiency and the infectious diseases. The dietary deficiencies can be remedied by feeding vitamins, essential amino acids, or the appropriate minerals. If the deficiency is an internal one, such as an insufficient supply of hormones, the diminished production can, in many cases, be supplemented by extracts of the organs of other animals or by synthetic preparations. The infectious diseases are being mastered with antibiotics and other chemotherapeutic agents.

The static description of the molecular components of the cell could not suffice in understanding the mechanism of life. The tools available to the chemist up to 1932 proved inadequate for the exploration of the dynamic interactions within a living cell and for the study of interactions of the macromolecular components when removed from the cell. But help was on its way, for in 1932 Urey discovered deuterium, which proved to be a magic wand that opened the door to the innermost secrets locked in a living cell. As we said earlier, it was a student of Urey's, Dr. David Rittenberg, and his associates, who adapted the isotopic tools to the study of life. They wrote in their first scientific contribution in this area in *Science* of August 16, 1935: "The number of possible applications of this method appears to be almost unlimited." The prediction was amply borne out. During the era of biochemistry ushered in by the isotopes nothing less than the complete reorientation of our concepts on the biochemistry of the body was achieved. Our knowledge of the dynamic state of the body's constituents, the synthesis of complex compounds from small precursors, the uniformity of basic reactions in all living organisms all stem from intelligence gathered with isotopic tools.

The term molecular biology is bandied about a great deal lately and is extolled by some as a new discipline. Actually, molecular biology was initiated when isotopes were first used to probe the intimate mechanisms within the living cell. Prior to this we could but perform a post-mortem analysis on the components of dead cells. We were practising in those days what I like to call molecular anatomy. But the isotopes enabled us to observe molecular functions within living cells and we thus began the study of molecular physiology or molecular biology.

There were many who were skeptical of the Bergmann hypothesis for protein synthesis. Those with biological intuition felt that protein synthesis must require a more complex system for its achievement. It was apparent to some of us that, in the search for such a mechanism, at first all the components of the cell—not just one purified enzyme—must be used. Since no one expected to produce large amounts of protein initially, isotopes became an indispensable tool for the detection of the miniscule amounts of proteins newly minted in the test tube.

Carbon-14 can be incorporated by appropriate chemical manipulations into any one of the twenty amino acids and into specific sites of their structure. Or, the label can be introduced randomly into all amino acids by growing a plant in a closed atmosphere containing C^{14} carbon dioxide. The amino acids isolated from the protein of such a plant will all be radioactive. If such amino acids are offered to an intact organism its proteins will become radioactive and the extent of the radioactivity is a measure of newly synthesized proteins.

A number of investigators started almost simultaneously to

study the incorporation of radioactive amino acids into surviving tissue slices. The rationale for such an experiment is as follows. The amino acid penetrates into an intact cell and fits into its normal working machinery. Testimony to the incorporation of the amino acid into the proteins of the cell is provided by the radioactivity of the protein that is subsequently extracted. The advantage of such a system is that a cell with its structures and components intact is performing the protein synthesis—but it is isolated and, therefore, a number of external forces can be imposed upon it and their effects can be assessed. For example, can such a system continue to incorporate amino acids if its energy-yielding mechanism is crippled by the addition of a specific poison?

The investigator who asked the most penetrating questions on protein synthesis and came up with answers with the strongest ring of truth is Dr. Paul C. Zamecnik of Harvard University. Dr. Zamecnik, a physician who turned to research, gathered around him an unusually talented group of young investigators at the Huntington Memorial Hospital in Boston. (The caliber of an investigator's junior associates is one of the indices of his own quality. His mind, his personality, his whole being is exposed to the daily scrutiny of the most exacting judges, the younger colleagues who must use the association as a springboard for their careers.) Moreover, Zamecnik was particularly fortunate in having as a colleague on the staff Dr. Fritz Lipmann, an eminent biochemist who has had some of the most penetrating insights into molecular mechanisms. Dr. Lipmann had earlier deduced the source of energy in a living cell which drives the chemical processes that require energy. Every organism from the tiniest bacteriophage to the mighty whale has a reservoir of chemical energy in the form of a special molecule, adenosine triphosphate—abbreviated, ATP. This compound is a battery that can be repetitively charged and discharged with chemical energy. The energy is stored in two chemical bonds which can be cleaved by appropriate enzymes, and the energy of those bonds can be channeled to make new chemical bonds to fashion any compound the cell needs. Lipmann was one of those who was skeptical of the Bergmann hypothesis. He proposed that ATP provides the energy for the weaving of a protein molecule from its precursor amino acids. Zamecnik and his co-workers soon confirmed this. They found that if the system which generates energy in a cell is poisoned with a drug, dinitrophenol, protein synthesis as measured by the uptake of a radioactive amino acid ceases completely. One of Zamecnik's associates, Dr. Philip Siekevitz, extended the frontier by showing that disintegrated cells, too, can incorporate amino acids provided various components of the cell, including the energy-yielding mitochondria, are present in a relatively undamaged state.

Such a reassembly in a test tube of the components of disintegrated cells was made possible by the knowledge gathered by a group who were interested in anatomy at the subcellular level. The founder of this school was Albert Claude of the Rockefeller Institute, who, as so often happens in biological research, had something entirely different in mind when he initiated his search. Dr. Claude had become interested in seeing whether he could detect tumor-causing viruses, whose devastating effects are the only evidence of their presence after they infest healthy cells. He chose to work with the Rous Sarcoma virus, which causes tumors in chickens. In his search for the cryptic virus he developed methods of disintegrating cells and separating the various components by subjecting them to increasing gravitational forces in a centrifuge. The different fractions collected by this method were examined under an electron microscope and correlations were attempted with the components of whole cells. This work requires patience and visual imagination in equal measure. The organelles of a cell present entirely different images when they are clumped together in a mass and when they are integrated into the structure of a cell. The incidence of artifacts is high, errors of interpretation are inevitable. For example, it took considerable skill and imagination on the part of two investigators, Drs. George Palade and Philip Siekevitz, to identify isolated clumps of pellets as structural components of the cytoplasm. These particles, which are called ribosomes, are globular structures rich in both RNA and protein. Figure 8.1 is an electronmicrograph of a covey of ribosomes. Zamecnik soon found that the ribosomes were essential components of the protein-making apparatus; the radioactive amino acids in such cell-free systems were quickly concentrated into the proteins that were either integral parts of the ribosome or were adhering to it. The acidic particles

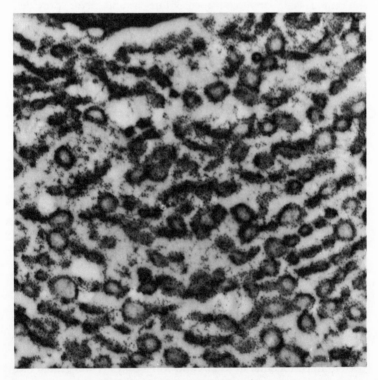

Fig. 8.1. Electron micrograph of ribosomes

which had been implicated in protein synthesis by Brachet and Caspersson were thus separated and identified at last as the ribosomes. However, the ribosomes alone were impotent in synthesizing protein. They had to be suspended in a soluble extract of the cells they came from, then, provided ATP was present as an energy source, the ribosomes could incorporate radioactive amino acids into proteins.

The reconstruction of components of disintegrated cells with sufficient finesse to achieve protein synthesis, however limited, was a signal achievement, although so far very little light had been shed on the intimate mechanism of the process. The "cell-free soluble extract," for example, may have hundreds of different, unrecognized components. Indeed, as it turned out, such extracts became gold mines of biochemical information. The miner who wielded his pickax most effectively was another associate of Zamecnik's, Dr. Mahlon Hoagland. He had been trained by Lipmann in methods of detecting the flow of chemical energy from ATP. From studies of simple models it became evident that ATP forms a transient chemical alliance with the substance which is destined to be fitted into a new molecular pattern. It is as if for the synthesis of complex molecules the precursors stand on the shoulder of ATP and thus climb the wall to higher levels of energy. Hoagland was able to show that there are enzymes in the "soluble extract" that achieve a chemical union between ATP and amino acids.

By this time, in 1955 to be exact, the impact of the Watson-Crick model of DNA was beginning to be felt; the role of nucleic acids in determining the sequence of amino acids was becoming part of the intellectual atmosphere. Zamecnik and Hoagland began to explore the role of RNA in protein synthesis: Is it just a passive template on which the protein molecule is stamped out or does the RNA of the ribosome serve a more active, a more involved function? To explore this question they designed a complicated experiment from which emerged an unexpected finding of paramount importance. The favorites of the goddess of serendipity are those who dare to leave the neatly laid out paths of yesterday's research and make bold forays into the thick jungle on the far horizon. Some of these ventursome souls become hopelessly lost, but those who are lucky and alert—"Chance favors the prepared mind," said Pasteur—may be rewarded by sudden visions of beauty, as hidden truth reveals the answers to the questions they are seeking.

An unexpected message from their Geiger counter spelled out one of these truths for Hoagland and Zamecnik. The question they were asking was whether the reconstituted cell-free system which could incorporate amino acids into protein might also incorporate nucleic acid precursors into RNA. If so, RNA would be made simultaneously with protein. They added to one half of their cell-free mixture the radioactive amino acid as before, and to the other half they offered a radioactive RNA precursor. A reagent called trichloroacetic acid—abbreviated TCA —can distinguish and separate RNA from protein. RNA dissolves in hot TCA, protein does not.

Therefore, if the "hot TCA soluble" fraction had been radioactive it would mean that RNA was being made from the nucleic acid precursor. They found this to be the case, but to their great surprise they found something totally unexpected. The "hot TCA soluble" material was radioactive even in the mixture to which the *amino acid* had been added. Evidently the amino acid was being channeled somehow into the RNA. The surprised investigators confirmed and extended their findings rapidly. They found that the amino acid was bound to a fraction of RNA which had previously been known to exist but whose function was totally obscure. But now a role for this RNA fraction had appeared at last. The amino acid was not yet bound into a protein but, rather, it was transiently associated with the RNA itself. The amino acid could be stripped away from the RNA by very gentle chemical manipulation. The RNA which accepted the amino acid, unlike the ribosomal RNA, does not sediment out even if it is subjected to 100,000 times normal gravitational pull. The name soluble RNA, or "s-RNA," was given to this material which Hoagland and Zamecnik had recognized as being the conveyor of the amino acid during the assembly of a protein molecule. It should be pointed out that biochemistry had advanced by then to a sufficiently sophisticated stage in its development that predictions of mechanisms could be offered, provided there was enough insight and imagination. That a carrier of amino acids probably exists had been predicted by two investigators, Crick and the Belgian biochemist H. Chantrenne.

An ingenious experiment revealed that soluble RNA is not a single substance but rather a complex one in which there are separate carriages for each of the twenty amino acids. This was revealed by observing that soluble RNA can be completely loaded with one amino acid until further addition of *that* amino acid does not increase the incorporation of radioactivity into the RNA. However, if a different amino acid is added, the previously loaded RNA accepts the additional burden. At this point still another amino acid could be loaded on.

Once the reality of soluble RNA was established other investigators flocked to its study, and now, ten years later, it is the nucleic acid about which we have the most detailed information.

Soluble RNA is relatively small, its molecular weight of about 25,000 is puny compared to that of DNA, which runs into the millions. The carriers of the different amino acids have been separated and were found to be indeed fastidious: Each amino acid has at least one separate soluble RNA at its disposal which takes it to the site of the erection of the protein. The portaging nucleic acid molecules contain about seventy-five to eighty bases. While the sequence of the bases in the different s-RNA's is as yet obscure, we know that the components at the terminal are the same. It is a sequence of two cytosines and one adenine. And we also know that the cradle into which the amino acid fits is a ribose molecule attached to the terminal adenine.

We can now write the first steps in protein synthesis in an abbreviated way as follows:

1. Amino acid + ATP + Enzyme → ATP ~ Amino acid
2. ATP ~ Amino acid + s-RNA-C-C-A

$$\rightarrow \text{s-RNA-C-C-A}$$
$$\wr$$
$$\text{Amino acid}$$

Or, if we prefer a visual image, every brick for the construction of an edifice of a protein molecule is brought in a separate wheelbarrow on a high platform and the brick is dumped onto the hod (the CCA end) of an individual hod carrier.

In this construction scheme the blueprint for the building of the protein molecule, of course, must be the DNA. But the DNA is in the nucleus of the cell and protein synthesis occurs on the ribosome, which is a component of the cytoplasm. How is the information for the ordered array of amino acids transferred from DNA to ribosome? The answer came to two investigators who were asking a different question. Drs. Elliot Volkin and L. Astrachan were studying in 1956 the effects of bacteriophage infection on the cellular economy of the infected bacteria. These workers at the Oak Ridge National Laboratories knew from earlier reports of others that in bacteria infected by a bacteriophage the synthesis of RNA essentially stops. But a certain amount of new RNA must be made because Volkin and Astrachan found that if radioactive phosphorus is offered to the bacteria right after the bacteriophage invades them, their RNA becomes radioactive. Since only the newly formed RNA contains the radioactive phosphorus, when the RNA is broken down to the component nucleotides—i.e., the base plus ribose phosphate—only the newly incorporated nucleotides will be radioactive. Therefore, the chemical determination of the amounts of the newly incorporated bases is easy; the amount of each base newly deposited is proportional to the radioactivity in the phosphate associated with it.

An arresting surprise emerged from these patient analyses. The RNA formed after the bacteriophage infection did not conform to the pattern of the four bases normally in the bacteria. Instead, the bases mimicked the ratio of the bases in the DNA of the invading bacteriophage. Volkin and Astrachan concluded that a "DNA-like RNA" was being produced as a result of the infection. In other words, the bacterial virus, after invading the host, disrupts the normal processes in the cell and somehow orders an RNA to be produced that is the mirror image of its own DNA.

How is such a "DNA-like RNA" made in the cell? In 1959 Dr. Sam Weiss, a young man working at the Argonne National Laboratories in Chicago, designed an experiment fashioned after Kornberg's successful synthesis of DNA in a test tube. Dr. Weiss took DNA and an extract from bacteria and added to it the four precursor nucleotides of RNA. (Kornberg, of course, had used the four building units of DNA.) Dr. Weiss was able to show that the building units were assembled into RNA. The assembly of the RNA was totally dependent on the presence of DNA, for if the latter was excluded or destroyed no RNA was produced. More than that, the newly fabricated RNA mimicked the base sequence of the DNA, which was used as the seed. Now, a "DNA-like RNA" was produced not only in bacterial cells invaded by bacteriophage but in a carefully reconstituted system in a test tube as well. Therefore, not only are there enzymes which achieve the duplication of DNA but there are still others which can translate the sequence of bases of DNA into the complementary bases of RNA.

At this point—in 1961—two highly imaginative French scientists, Drs. Francois Jacob and Jacques Monod, both of the Pasteur Institute, synthesized these disparate bits of information into a unified theory of the molecular mechanisms of the genetic apparatus. Part of their contribution was the coining of vivid phrases for the processes and products involved so that ideas could be easily verbalized and exchanged.

The first step in the mobilization of a gene into action is the "transcription" of the sequence of bases in the DNA into a complementary sequence of those bases in RNA. The RNA to which the genetic information is entrusted is a "messenger" RNA. In the shaping of the messenger RNA the enzyme is supposed to copy only one strand of the DNA. The alignment of the bases of the RNA is presumably by hydrogen bonding with the appropriate base pair of DNA. The one cardinal difference from base-pair alignments in DNA is that, in forming RNA, uracil substitutes for thymine in pairing with adenine as seen below.

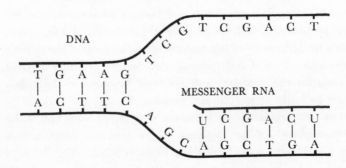

RNA Ribose
Uracil   Deoxyribose DNA
Adenine

The "DNA-like RNA" which forms in bacteriophage-infected cells was renamed as "messenger RNA." With the aid of some exquisitely sensitive methods of centrifugations developed by three exceptionally ingenious physical chemists, Drs. J. Vinograd, Richard B. Roberts, and Roy J. Britten, the messenger RNA from such infected cells could actually be separated from other RNA's and its properties studied. The earlier conclusions of Volkin and Astrachan, which were drawn from studies of mixtures of all the RNA's, were now confirmed with isolated samples of messenger RNA. A scheme for the beginning of the transcription of one strand of DNA into a messenger RNA is presented below.

How is the message in messenger RNA expressed? Or, to take the questions back one step further, how is the information for a sequence of amino acids for a protein encoded in the DNA? As soon as it became firmly established that DNA is the repository of such information, the paradox of the paucity of bases versus the abundance of amino acids confronted us. As many as twenty different amino acids may appear in a protein, but there are only four component bases in DNA. Obviously the only way to stretch a four-letter code to become an alphabet for twenty amino acids is by the use of some multiple of those bases for a single amino acid. The first one to play with such a numbers game in print (in 1952) is the American biochemist A. L. Dounce. If it is assumed that two bases spell out one amino acid, the number of permutations possible, 4^2 or sixteen, still falls short of the need. The assumption that a sequence of three bases represents a given amino acid provides a harvest of 4^3 or sixty-four permutations which is, of course, more than adequate. This is a rather arbitrary numerological approach to a biological problem. We cannot rule out the possibility of four bases to a code, and assume that only twenty of these possible 256 (4^4) have survived selection through eons of evolution. But for the time being let us accept the three-base code, which we shall discuss in greater detail in the next chapter.

We have as yet no idea how the enzyme which astutely transcribes but one strand of the DNA into messenger RNA makes its choice. Nor do we know what serves as a punctuation mark in the DNA to indicate the start and the end of the specific gene that is to be copied. Whatever these working details of the transcription are, once the messenger RNA is completed, it is supposed to peel off and migrate—propelled by we know not what force—into the cytoplasm. There it attaches itself to a ribosome and awaits the s-RNA's, each of which bears its distinctive load of a particular amino acid.

How does the s-RNA recognize the site where it is to bring its amino acid so that the next operation, the "translation" of the message in the messenger RNA into a sequence of amino acids, can be unerringly performed? This part of the hypothesis on the mechanism of protein synthesis is the most nebulous and has the least support from experimental facts. A sequence of three bases in the s-RNA is presumably the complement of the three-base code word in messenger RNA. For example, let us assume for the time being that the code in messenger RNA for the amino acid glycine consists of the three bases, uracil, guanine, guanine, abbreviated UGG. The complementary recognition triplet in s-RNA would have to be adenine, cytosine, cytosine. In this way the UGG of messenger RNA could pair off with ACC of the s-RNA.

Now the next s-RNA, let us say the one bearing alanine, would move into the adjacent site, beckoned there by the appropriate three-base permutation which codes for that amino acid.

The amino group of alanine would now shove itself into the area of the electronic bonds connecting glycine to its s-RNA, and a new coupling between glycine and alanine would be formed. Such transfer of bonds is perfectly feasible; actually there are many counterparts of it in organic chemistry. A third amino acid on its s-RNA would now approach alanine and fuse with *it.* So the process of stepwise accretion would continue until all the information of the messenger RNA is translated into a completed protein molecule. In Figure 8.2 a bird's-eye view of the process of protein synthesis is represented. According to our current views, the messenger RNA now might move to another ribosome and there the same process of translation could ensue. Eventually the messenger RNA would be cleaved by enzymes and its message thus destroyed. The ribosomes would be ready to welcome a new messenger RNA and to shape a different protein molecule.

The scheme of protein synthesis I have just described is probably valid in its large outline—but undoubtedly deficient in most of its detail. A cosmonaut from outer space, viewing our planet

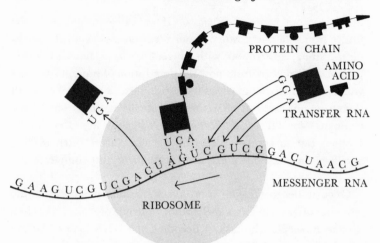

Fig. 8.2. Schematic view of protein synthesis

from his space ship several hundred miles away, would probably conclude rightly that our Earth is cultivated and inhabited. However, he could not even guess at the structure of our family, society, or nation. Our view of protein synthesis is a cosmonaut's view. Parts of the hypothesis about areas as yet unpenetrated and unconfirmed by experimental probes lack the ring of truth to some of us. For example, the recognition by the s-RNA of its site on the messenger RNA puts too much of an imposition on our chemical credulity. It is relatively easy to accept the alignment of single bases by hydrogen bonding during the synthesis of DNA or messenger RNA. These bases are of low molecular weight and the hydrogen bonds are adequate to secure them. But in the alignment of s-RNA on the messenger RNA we are proposing that a large structure with a molecular weight of 25,000 gropes its way along until three out of its seventy-five bases mesh with the right code word in the messenger. The speed with which the alignment and the zipping together of the amino acids occur point to the handiwork of enzymes, which are yet to be discovered and fitted into our current concepts. If we can extrapolate to the future from the past history of our ideas on biochemical mechanisms, smug satisfaction with current views is fatal to future progress. Ten years from now the description of protein synthesis will be both more simple and more complex. The simplicity will stem from our recognition of the intimate mechanisms which now elude us—for all basic biochemical reactions are disarmingly simple. But to achieve such simple tasks Nature often evolves ancillary mechanisms of exquisite refinement and complexity.

Once it became clear from Avery's achievement that some sequence of bases in DNA spells out a sequence of amino acids in a protein molecule, the task of decoding the DNA loomed before us. It is an unfailing attribute of sound research that every answered question yields a harvest of new questions. With each sure step upward a larger horizon unfolds revealing newly shaped concepts and newly formulated tasks for further exploration. And it is a qualifying attribute of a real scientist that he accepts the new challenge. However, the task of decoding DNA staggered even the bravest among us. Consider the problem. We can make some quite accurate approximations on the number of different protein molecules in a cell; in turn we can estimate the minimum number of code letters of DNA needed to encompass all that information. Someone has calculated that for the storage of information for the proteins of the invisible colon bacillus—assuming a three-base code—a closely typed book of 3000 pages would be needed. The only apparent hope of achieving a decoding was to await the time when we may isolate in pure form a single gene which has inscribed in it the information for the weaving of a protein whose amino acid sequence is known. But since, as yet, we have not isolated a gene in pure form, nor have we been able to make any headway in determining base sequences in DNA, the accomplishment of decoding seemed decades away. But the biological scientist must strive not to be stayed by defeatist pessimism. Those who daily go to their laboratories to search and find things never seen or understood before must have hope. Without such hope we might as well stay home and "create" nonobjective art.

By 1961 many investigators were studying in detail the cell-free, protein-synthesizing system which Zamecnik and his associates had perfected. One of the outstanding laboratories where the relation of RNA to protein synthesis was being explored was at the National Institutes of Health in Bethesda. These institutes— which are part of the U.S. Public Health Service—have a proved history of excellence. Their scientific staffs are second to none; their cumulative achievements are truly impressive. For an institution under the close scrutiny of Congress, the atmosphere is surprisingly relaxed. The drawbacks inherent in a government position are compensated for by adequate space, funds and facilities for research. Rivalry and competition seem to be at a minimum. Everyone knows everyone else's salary, which is determined by Congress; no dean has to be impressed or cajoled for special preferment.

A young man with a rank of G(overnment) S(ervice) 12 was working there. A G.S. 12 is rather a low man on the totem pole of govenment-employed scientists. His salary range currently is ten to twelve thousand dollars. As it turned out the salary paid to the young man was the best investment—dollar for value—our government has ever made, including the purchase of Louisiana and Alaska.

"Wondrous the Gods, more wondrous are the men."

BLAKE

9 THE CODE OF LIFE

A RIPPLE of excitement waved through the participants of the Fifth International Congress of Biochemistry in Moscow in 1961. A new finding described at one of the many symposia was nothing less than sensational: The first letter of the genetic alphabet was deciphered; the code was cracked.

The audience which first heard the report of one of the most important biological experiments of the century numbered less than a score.

But such sessions in which supposedly new findings are briefly described are usually the least important ingredients of international scientific meetings. The organization of the meetings is slow, the reports must be submitted at least six months in advance, and, therefore, very seldom is anything truly new unveiled at these international gatherings. However, there are other satisfactions both professional and personal which draw together, from all over the world, the communicants of a given discipline. The symposia, which are obligatory ingredients of such meetings, are the most rewarding experience professionally.

Since adequate time is given to each speaker who has presumably made a significant contribution to his field, one can become acquainted with the concepts and achievements which have been made in areas other than one's own. For the sad truth is, written communication among scientists has broken down. This has occurred not in spite of but because of the enormous volume of scientific literature which deluges us.

At the start of the era of science, exchange of information was informal and simple: those who had the interest, leisure, and the means to engage in scientific pursuits would write to each other describing their findings and conclusions. It was not unusual for the recipient of such information to report it to some scientific body in his country. Since there were no rewards for scientific pursuits anyway, no fear of loss of priority gnawed at the vitals of these blessed amateurs.

This happy informality of course vanished as natural philosophy became fragmented into the various branches of science and as the number of practitioners burgeoned.

Specialized journals were founded and discoveries began to trickle in from all over the world. Over the years the trickles swelled to rivers, but one could still remain afloat. In my field, biochemistry, one could be well informed twenty years ago by skimming a half a dozen journals a month, and reading with care the few articles which were close to one's special interest.

But now, since the scientific professions have experienced a population explosion of their own—someone has calculated that 80 to 90 percent of the scientists who ever lived are alive today—we are drowned in the cataracts of journals pouring on us relentlessly from all over the world. Moreover, since so much is being published, the journals insist on a telegraphic style built around abbreviations and a specialized jargon. The contents of the journals have become arcane codes comprehensible only to a small coterie of experts in a tiny area. (The increasing disappearance of skill in writing a clear sentence from the arsenal of equipment of otherwise well-trained people does not help any either.)

As communication of ideas through the journals had become increasingly frustrating we have searched for new solutions and we have made a profound new discovery: the best way to achieve an exchange of ideas is by bringing people together.

So we are back where science started: we are encouraging personal communication between scientists. Every center of research has small group meetings and seminars by the dozen each week. In most large cities there are monthly dinner meetings bringing together people of mutual interest from a locale of larger radius. Each subdiscipline of science holds at least one annual national and usually a triennial international meeting. But even all of this proved inadequate. The national and international meetings tend to be huge and formal and lack the facilities for relaxed gathering of those working in the same small area. Moreover, these large meetings are by administrative necessity organized by the appropriate professional societies for their own members and therefore nonmembers tend to be excluded. This is a shortcoming because as we need more and more sophisticated tools and techniques to explore the complex world around us, participation in the same area by workers with different technical backgrounds must continue to increase. In biology especially, it is not unusual for physicists, chemists, and bacteriologists to be working on the same problem. To afford an opportunity for workers with a specific goal to meet beyond the confines of their professional denominations and to provide maximum opportunity for unfettered conversations, small international meetings are arranged to focus on special topics. The host country sends invitations to a small number of workers all over the world. The sites of the meetings, of course, tend to be typical of the hosts and thus offer a startling variety. The French will take you to a thirteenth-century abbey, the Israelis isolate you on the shores of the Sea of Galilee, and we might hold such a meeting at a ranch in Estes Park. Isolation and communal living is a routine requirement; conversations at the breakfast table or on strolls around the compound are just as important and fruitful as the organized conferences.

The choice of the quarters for the meeting in Moscow, the building of the Lomonosov University, was unfortunate. This is one of about half a dozen almost identical structures which dot the skyscape of that drab city. They are ghastly monuments by means of which Mr. Stalin set out to prove that communism can compete successfully with capitalism in the building of ugly monstrosities. All of these buildings, whether they house a university, or the bureaucracy for agriculture, or serve as a hotel, are almost identical in appearance. Some wag at the meeting dubbed them Stalin Gothic. They are enlarged and distorted versions of the Wrigley Building in Chicago. From a massive double-winged pedestal about twenty stories high a tower lumbers toward the sky. The Chicago prototype of these is hardly the most graceful or impressive example of skyscraper architecture, but even its modicum of appeal—the whiteness, the filigree effect of its stonework—was lost in translation. The Moscow structures have a drab, sandstone color and their ornamental stonework is gargantuan. (Grecian urns two stories high were the original devices created by some inspired Slavic soul to ornament the Hotel Ukraine.)

The buildings were designed without regard to their human inhabitants. There were 6000 participants of the Congress, and the meeting rooms were mostly on the lower floors. The crowding in the inadequate corridors equaled that in the New York subways at rush hour.

Under these circumstances it is little wonder that only a handful were present when a young, obscure biochemist, Dr. Marshall Nirenberg, did his ten-minute stint. There was no inkling in the brief printed abstract of his paper of what he was to report, for he made his discovery after the abstract had been submitted. Therefore, only Dr. Nirenberg knew what he was about to say, only he knew its import. Fortunately, one of the more gifted young molecular biologists, Dr. Matthew Meselson, was in the audience and he, too, recognized the worth of the finding. Dr. Meselson reported what he heard to Dr. Crick. The reaction of that versatile, unorthodox genius was worthy of him. The reception of a truly new idea from an obscure source is a test of the recipient. The easiest reaction is disbelief in its reality, the safest is skepticism of its meaning. Statistics favor the skeptic. After all, of the thousands of annual communications in the field of biochemistry, not a few are blunders, many are artifacts, most are trivial extensions of previous knowledge. But a handful of such reports contain new and occasionally exciting revelations about the workings of a living cell. Such a revelation must struggle for recognition against the entrenched skeptics whose number is legion. Skepticism is the spawn of small and timid minds; faith in others shines only from bold and bright intellects who, first of all, have faith in themselves.

Dr. Crick invited Nirenberg to repeat the delivery of his paper before the large audience of a symposium of which Crick was the chairman. Such a rapid recognition of a contribution from an obscure young scientist is unparalleled in the history of science. It was thus that many of us in Moscow had a chance to hear the tall, handsome, modest but strong young man from the National Institutes of Health at Bethesda.

In order to be able to understand Dr. Nirenberg's electrifying announcement it is essential to acquaint the reader with the base on which the young man built his achievement.

Dr. Nirenberg was one of the young men at the National Institutes of Health who was interested in the relationship between nucleic acids and protein synthesis. He is a perfectionist in his approach and he was polishing the components of the system which can approximate protein synthesis in a test tube. The reader may recall from the previous chapter that to achieve such a goal we use the components of disintegrated cells. These include ribosomes, enzymes, transfer RNA, ATP as a source of energy for the reaction, and preparations of other RNA, which we now call the messenger.

With patience and skill Nirenberg secured the ground for obtaining consistent results from these complex concoctions. Often a very simple technical improvement can speed the progress of a project. For example, Nirenberg found that storage of his enzyme preparations at $-80°C$ instead of $-15°$ preserved them almost indefinitely. This sounds like a picayune contribution, but it ensured the availability of uniform preparations of one component of his system. Thus the effect of other, more variable additives could be confidently assessed.

Nirenberg began to test the effect of various RNA's on the incorporation of amino acids into proteins. He obtained a preparation of tobacco mosaic virus RNA and found this to be a very active agent for aligning amino acids into proteins. He went on to test other RNA's. One of these has a curious history. A lady biochemist from Paris, Dr. Marianne Grunberg-Manago, came to New York to Dr. Severo Ochoa's laboratory on a travel fellowship in 1955. She was asked to study the metabolic fate of a compound containing adenine, a ribose, and two phosphate groups. This is adenosinediphosphate. We might abbreviate this as ADP or symbolize it as

She exposed ADP to extracts from bacteria and looked for alterations that may have occurred to it. This was done by paper chromatography of the products that may have resulted from exposure to the bacterial juices. To her great surprise some of the ADP refused to migrate upward in the usual manner: part of the material stubbornly remained at the starting point. This is the sort of chance observation which may be made by many who work in the complex world of experimental biology. But the reception of the chance phenomenon varies with the observer. His reaction is a test of his intellect, imagination, and inner strength. Many reject such observations—especially if it is reported by an underling. I know of one anomalous observation made by a technician which, had it been correctly interpreted, would have become an outstanding discovery. But the young man was rebuked for his inability to obtain the anticipated result and was fired. Unfortunately, there are all too many workers with such rigid attitudes which ultimately stem from their lack of imagination. Many of them sit in their laboratories with the same probability of producing anything really new as a hen sitting on a clutch of unfertilized eggs. These are the people who complain of the pressure to "publish or perish." Given the vast complexity of the living cell and our present limited knowledge of it, anyone who cannot make novel observations is like an explorer who, say a hundred years after Columbus's discovery of America, could find no new lakes or mountain ranges here. They both should stay home.

Dr. Grunberg-Manago recognized the meaning of the sluggishness of her products during paper chromatography. This is the characteristic behavior of large molecules, not of the small compound with which she started. The unitary molecule ADP

A
| P
| /   must have been converted by some enzyme into a
| P
| /
|/

large aggregate which might be symbolized as

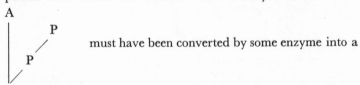

Physico-chemical analysis of the attributes of the product indeed confirmed such an interpretation: She had produced an RNA-like product which contains only one of the four usual component bases. Dr. Severo Ochoa, in whose laboratory the discovery was made, at first thought that this may be the method of synthesis of RNA and therefore developed the discovery with vigor. It was found that such homogeneous macromolecules which simulate RNA can be made with any of the four bases. The availability of these products proved to be highly profitable. For example, they permitted a test of the Watson-Crick hypothesis. The reader will recall—from Chapter 7—that according to this hypothesis the base adenine becomes closely associated with the base thymine to form the double helix. We could now make a homogeneous RNA-like polymer which contains nothing but adenine abbreviated as "poly A," or thymine abbreviated as "poly T." Since pure solutions of the two species of macromolecules were now available we could study what happens when the solutions are mixed and the two artificial RNA's are permitted to co-mingle. Anyone who has seen a car filled with boys and a car filled with girls and a car filled with boys *and* girls can predict what will happen when "poly A" and "poly T" are mixed. The heterogeneous system—be it boys and girls or the artificial RNA's—occupy less space than the homogeneous ones. In Figure 9.1 there is a schematic representation of such fusion of poly A and poly T via hydrogen bonding.

The enzyme—for such it proved to be—which clips together the small molecular weight components into the large complexes is an unusually undiscriminating one. Rather lackadaisically it weaves together whatever is offered to it. If only one base precursor is available it will make a homogeneous polymer of that one; if two, three, or four base precursors are offered simultaneously it will make polymers containing whichever components were accessible and approximately in the ratio in which they are present initially. It was obvious almost from the start that

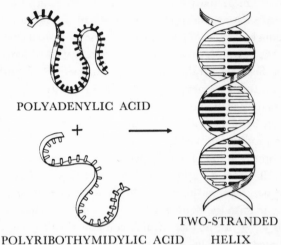

POLYADENYLIC ACID

+

POLYRIBOTHYMIDYLIC ACID

TWO-STRANDED
HELIX

DR. A. RICH

Fig. 9.1. The birth of a double helix

this mechanism cannot be the one that makes real biological RNA. The exquisite individuality of a component of the cell, such as RNA, could not be preserved if its structure were permitted to be determined by the relative amounts of building units which happen by chance to be present at any one time. Such a scheme of RNA synthesis harks back to the Bergmann hypothesis of protein synthesis by the reversal of degradative enzymes. There, too, the abundance and variety of amino acids were to determine the structure of the protein.

The discovery in 1959 by Weiss of enzymes which made RNA only in the presence of DNA and which faithfully copied the structure of that seminal substance dispelled completely any credence in the enzyme discovered by Dr. Grunberg-Manago and studied intensively by Dr. Ochoa as having a role in RNA synthesis. To this date we are uncertain about its real function. An educated guess is that it destroys the messenger RNA so that translation of it into protein cannot continue indefinitely. It resembles the paper chopper in a secret coding room which, at the end of a day, snips into shreds all of the tapes which bear any information. However, a stellar role was awaiting, not this enzyme, but one of its products.

Among the variety of RNA's tested by Nirenberg as messenger RNA was a homogeneous RNA-like product which contained only uracil. Its structure might be schematically represented as

Its laboratory nickname is "poly U."

The outcome of this experiment is so momentous the details bear repeating. Nirenberg and his associate, Dr. J. H. Matthaei,[1] assembled the following ingredients: ribosomes, s-RNA, and the enzymes which attach amino acids to s-RNA. The energy source which had been perfected by Zamecnik's group was obtained from commercial sources. Twenty amino acids which had been isolated from a plant that had been grown in an atmosphere of C^{14} carbon dioxide—and as a result of which all of the amino acids were radioactive—were also commercial products. When these ingredients, ribosomes, s-RNA, enzymes, energy source, and radioactive amino acids were mixed, the incorporation of amino acids into protein was practically nil. In Figure 9.2 the low, horizontal line of the left panel indicates the kind of negative results which were obtained under these conditions. When "poly U" was also added to the mixture the incorporation of radioactivity, therefore, of one or more amino acids was enormous. If to 1 cc of the incubation mixture one hundred thousandths of a gram (10^{-5} g) of poly U is added the incorporation is stimulated a thousandfold. The upper, ascending curve in

[1] Dr. Matthaei was in Nirenberg's laboratory as a postdoctoral fellow from Germany. His stay was financed by, of all organizations, the NATO! A noteworthy instance of swords into plowshares.

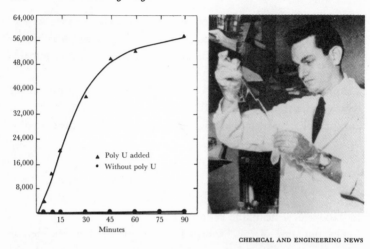

Fig. 9.2. Dr. Marshall Nirenberg at work

Figure 9.2 represents the results of such an experiment. When the startled investigators recovered from their astonishment they asked the obvious question. Does "poly U" stimulate the incorporation of all, or some, or one of the twenty amino acids. The answer to this question can be had in six or five experiments, depending on the gambling instincts of the investigator. Four different groups, each containing five different radioactive amino acids, were assembled. When these four groups were tested only one could yield radioactive protein under the directive influence of poly U. The five amino acids of that group could be tested separately or in two groups of two and three and then the active group subdivided into two or three individual ones. Finally, it was found that only one amino acid, phenylalanine, was incorporated into a protein like material under the directive influence of poly U. Chemical analysis revealed that the radioactive material was a high molecular weight product which contained only phenylalanine. The mechanism now became clear. Some multiple of the base, uracil—it may be three, four or more—in

RNA spells out the information for phenylalanine to be inserted into a protein. Since poly U is nothing but a long sequence of uracils, nothing else is encoded in it. Consequently it repeats its message like a broken record: "phenylalanine, phenylalanine, phenylalanine . . ." and therefore the polymer containing only phenylalanine accumulates.

The code was cracked, the first letter of the genetic alphabet in RNA was deciphered:

$$\text{RNA UUU} = \text{Protein Phenylalanine}$$

The number of uracils is arbitrarily assumed to be three for the time being.[2]

The reaction to the news was revealing. Biochemists with abundantly documented past history of lack of imagination sneered. On the other hand, I recall telling the story a couple of days after returning from Moscow to someone who has made brilliantly original contributions to a field distant from Nirenberg's achievement. He said: "Then we are right about the

[2] Rumors abound that Dr. Nirenberg did not design to use artificial RNA's as possible messengers, but stumbled on his finding fortuitously. Even if this were the case, it would not detract from the discovery; but there are sources of reliable evidence testifying that those experiments were part of a systematic program designed by the young man. A historian should not dignify scuttlebutt by according it printed permanence, but denigrating rumors are part of a pattern of the reception of new ideas by all too many in the profession. When anything really novel is stated the first reaction on the part of many is that it is not true. Once the reality of the finding is proved, the next accusation is that it is not new. Finally, when these sullying attempts fail, comes the last line of self defense by the skeptic: the discovery was made accidentally by a technician. The reason for such last-ditch resistance to novel concepts or findings is very simple. It is inconceivable to a mind lacking any imagination that anyone else can have a creative imagination. This is not to remove a candlepower from Dr. Nirenberg's brilliantly illuminating achievement, but the field was ready for such a bright and bold experiment. The writer is aware of an authenticated case of another young man who wanted to use synthetic RNA's to probe the genetic code. Unfortunately, in his case his senior supervisors dissuaded him from trying anything so obviously futile. Boldness wins the day not only in love and war: Nirenberg is known in every corner of the world where there is a biochemist; the other young man must, for obvious reasons, remain anonymous.

whole business of DNA, RNA, protein." And indeed this was the first, most important impact of Nirenberg's discovery. Seventeen years after Avery's announcement that DNA is the genetic material, the last link of evidence for the cyclic mechanism of life

was forged.

For Nirenberg the first triumph was a clear signal—more work. At lunch in Moscow he outlined his plans to a friend.[3]

Since a multiple of uracil in the synthetic RNA is the code word for phenylalanine, the search for the code words for other amino acids should be relatively simple. Assuming a three-letter code, the four bases in DNA, adenine (A), cytosine (C), guanine (G), and thymine (T), can form 64 different permutations.

AAA	GGA, GAG, AGG
CCC	GGC, GCG, CGG
GGG	GGT, GTG, TGG
TTT	TTA, TAT, ATT
AAC, ACA, CAA	TTC, TCT, CTT
AAG, AGA, GAA	TTG, TGT, GTT
AAT, ATA, TAA	ACG, AGC, CAG, CGA, GAC, GCA
CCA, CAC, ACC	ACT, ATC, CAT, CTA, TAC, TCA
CCG, CGC, GCC	AGT, ATG, GAT, GTA, TAG, TGA
CCT, CTC, TCC	CGT, CTG, GCT, GTC, TCG, TGC

[3] And there was plenty of time during meals for such discussions. The Russians are a marvelous people and their technical progress since 1917 is nothing less than prodigious. In a little over two scores of years they have risen from their mire of feudalism to a level of technical sophistication which enables them to send a man to outer space and back. But the task of sending an individual meal from the kitchen to the dining room still baffles them.

Messenger RNA presumably contains the complementary base sequence of DNA. For example, if we take at random, the code ACA in the DNA dictionary, its RNA complement would be UGU. An RNA-like polymer containing two parts of uracil and one part of quanine is relatively easy to make.

One mixes two parts of uracil ribosediphosphate and one part of guanine ribosediphosphate and offers the mixture to Dr. Grunberg-Manago's enzyme and the dull-witted but agile-handed enzyme goes to work. The product may be

or $(UGU)_x$.

Such an artificial RNA can be added to an amino acid incorporating system and each of twenty radioactive amino acids can be tested individually. If one of the amino acids is preferentially incorporated then the UGU of the RNA is the code word for that particular amino acid.

With the path clearly pointed by Nirenberg's initial discovery, progress was rapid. The administration of the National Institutes of Health wisely put a larger staff and increased facilities at his disposal enabling him to tool-up to an intensive code-breaking operation. Moreover, he soon had company. Dr. Ochoa in New York and Dr. Grunberg-Manago in Paris used their vast experience with the enzyme to prepare scores of combinations of the four bases into artificial RNA's. The different preparations were tested patiently in several laboratories with the amino acid incorporating system perfected by Nirenberg. Confirmation of the validity of the approach rapidly emerged: specific premutations of the bases stimulated the incorporation of particular amino acids. A code-word dictionary of RNA bases specifying every amino acid was rapidly compiled.

A fastidious specificity was revealed by the data. For example, glutamic acid appears in proteins in two different forms, one with both acidic groups free and one with a lid of ammonia on it. The latter is called glutamine.

$$
\begin{array}{cc}
\text{COOH} & \text{COOH} \\
| & | \\
\text{HC—NH}_2 & \text{HC—NH}_2 \\
| & | \\
\text{HC—H} & \text{HC—H} \\
| & | \\
\text{HC—H} & \text{HC—H} \\
| & | \\
\text{COOH} & \text{CO·NH}_2 \\
\textit{Glutamic Acid} & \textit{Glutamine}
\end{array}
$$

Each of these compounds is specified by a different code word in RNA—and therefore in DNA. The selectivity between the two forms of the amino acids starts to operate at the recognition by an enzyme which attaches only one, either glutamic acid or glutamine, to its unique soluble RNA. In turn, recognition of the code-word site in the messenger RNA is achieved not by the amino acid itself but by its carrier, the soluble RNA. We have concluded this from an ingenious experiment devised and executed by a consortium of investigators: Drs. Lipmann and Chapeville of the Rockefeller Institute, Dr. von Ehrenstein of Johns Hopkins, and Drs. Benzer, Weisblum, and Ray of Purdue. The scheme depends on the interconvertibility of the two amino acids cysteine and alanine by chemical means. If a sulfur atom is removed from cysteine it becomes alanine. (A hydrogen atom covers the shorn site where the sulfur had been.)

$$
\begin{array}{cc}
\text{H} & \\
| & \\
\text{S} & \text{H} \\
| & | \\
\text{H—C—H} & \text{H—C—H} \\
| & | \\
\text{H—C—NH}_2 & \text{H—C—NH}_2 \\
| & | \\
\text{COOH} & \text{COOH} \\
\textit{Cysteine} & \textit{Alanine}
\end{array}
$$

This transmutation can be performed *after* cysteine is attached to its individual soluble RNA. Essentially we now have the amino acid alanine adopted by a foster soluble RNA whose normal attachment is cysteine. The question is which RNA code word, the one that specifies alanine or the one that specifies cysteine, will orient the incorporation of the amino acid from this foster carrier.

As it turned out, it was the code word for cysteine which opened the processes of incorporation of the adopted alanine. Recognition of sites on messenger RNA where the amino acid is to be taken is therefore achieved by the RNA which bears the amino acid.

With the help of the Nirenberg Code we could glimpse some of the hidden mechanisms of protein synthesis but, as often happens in experimental biology, the initial apparent simplicity was drowned in the gush of rapidly accumulating new evidence. It soon became apparent that more than one permutation of bases could direct the incorporation of the same amino acid. The number of code words accumulated is actually 64.

The ugly expression "degenerate code" was borrowed from the cryptographer's terminology to describe the multiplicity of RNA code words. Some of us were worried about this aspect of the conclusion about the genetic code. A degenerate code is devised by a cryptographer to impede unauthorized deciphering by creating confusion. All other biochemical mechanisms open to our view exhibit exquisite specificity to avoid any possibility of confusion. However, the advantage gained for protein synthesis by a relaxation of such rigid biological specificity has become clear. Purification of the soluble RNAs which transport the amino acids revealed that in several instances there is more than one RNA which can carry the same amino acid. Thus, there are two different soluble or, to use its more recent name, transfer RNAs, which can accept the amino acid, methionine. One of these transfer RNAs responds to the code AUG in messenger RNA. The other to GUG. The need for these two different transfer RNAs became clear from work in Dr. Sanger's laboratory. (Yes, it is the same Dr. Sanger who decoded the structure of insulin. The adage, "Lightning does not strike in the same place twice" does not apply to scientists. However, this is, of course, self-generated "lightning.")

Methionine attached to one of these transfer RNAs is sought out by an enzyme which then produces a chemical alteration of its amino group making it incapable of receiving a coupling from the acidic group of another amino acid. Therefore, the altered methionine attached to that transfer RNA can only react through its *own* acidic group. In this condition it is an ideal agent for *initiating* a protein chain. Thus, this unique transfer RNA which responds to the Code GUG is a chain initiator and responds only to its code which must be at the beginning of a "message" in messenger RNA.

Should this transfer RNA, bearing its altered methionine get into the middle of a message it would abort the production of that particular protein for no attachment could be made to the altered amino group.

There are telling lines of evidence attesting the over-all validity of the Nirenberg Code. The reader will recall that we know the defect in sickle-cell hemoglobin: a glutamic acid was replaced, due to a genetic blunder, by the amino acid valine.

What had occurred in the RNA to bring about the unfortunate shift: glutamic acid   valine? The code word in RNA deduced for glutamic acid is GAA; for valine it is GUA. It is apparent that the minimal change of the substitution of a U for an A in RNA will transform

$$GAA \longrightarrow GUA \text{ and, consequently,}$$
$$\text{Glutamic Acid} \longrightarrow \text{Valine}$$

The original mutation which must have occurred in the DNA

of some ancestor of the millions who today suffer from sickle-cell anemia can be symbolized the following way:

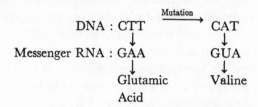

(The CTT and CAT are the sequences in DNA complementary to the GAA and GUA of the messenger RNA.)

There are a host of other known mutations involving amino acid exchanges which also can be accounted for by single base changes in the code words assigned to them. The most impressive series of these come from the laboratory of Dr. Charles Yanofsky, a brilliant biologist whose efforts, completely different from Nirenberg's, might eventually have led to the same goal, the deciphering of the genetic code. Yanofsky has been studying a large series of mutants of *Escherichia coli* which cannot make the amino acid tryptophane. Their deficiency is due to their inability to make an enzyme active in the assembly of this amino acid. Yanofsky and his associates found that in many of these mutants the enzyme is present but is inactive because of some structural alteration. These enfeebling alterations turned out to be amino acid exchanges which Yanofsky patiently determined.

In every instance these mutations—they are called point mutations because of their narrow range—can be accounted for by a single base change in the assigned code word. (From the laws of probability we should expect single base changes to occur with the greatest frequency.) We must look upon these amino acid changes and the base alterations to which they are ascribed as powerful arguments for the validity of the code words deduced for the amino acids. Since Nirenberg's code withstood the kind of analysis provided by the above mutants it will probably endure further scrutiny. We may anticipate minor alterations in detail but no gross changes in their fundamental pattern. The deciphering of the code will stand as a monument to human ingenuity.

But lest we become besotted with too much pride, we should remind ourselves that ultimately we owe it all to DNA. Somewhere within the labyrinthine structure of that wondrous stuff is invested the potential for the development of a brain capable of decoding the master molecule, DNA itself.

What is the practical value of all this? Given the information that sickle-cell anemia is caused by a shift of thymine to adenine in the gene which is responsible for the synthesis of hemoglobin, can we hope to remedy this error and return the millions debilitated by it to normality, within the near future? Unfortunately, the only immediate hope is for palliative measures. In galactosemia, a bypassing of the deficiency, by keeping milk from the infant, is perfectly adequate. In sickle-cell anemia our knowledge of the genetic change offers no easy therapeutic guide. For the time being, the best we can hope for is the perfection of some drug which might specifically bind to hemoglobin, bringing with it the lacking negative charge. At the present time our skills and tools are too crude for interceding with the structure of DNA with sufficient finesse to remedy a human genetic deficiency in a particular individual.

But what lies ahead? The details no one can predict; they are stacked neatly in the storehouse of the future. But of this we can be certain: generations yet unborn will wield their ever increasingly discerning tools and build monuments on the foundation of today's lasting achievements. New heights will be reached, new splendors revealed, and untold bounties for human welfare harvested. The unpredictability of the future of science in the next fifty years may be gleaned from a backward glance over our shoulders to fifty years ago. In 1916 the English biologist William Bateson had this to say about the possibility of exploring the molecular mechanism of heredity.

It is inconceivable that particles of chromatin or of any other substance, however complex, can possess those powers which must be assigned in our factors (genes). The supposition that particles of chromatin, indistinguishable from each other and almost homogeneous under any known test, can by their material nature confer all the properties of life surpasses the range of even the most convinced materialism.

But the efforts of the past fifty years belied Bateson's pessimism rooted in vitalism. We see "chromatin" today, as the distinguished biochemist Dr. Thomas H. Jukes put it, "elementary in the simplicity of its fundamental structure, yet labyrinthine in its complexity, ageless in its continuity, and infinite in its variety." And as we have seen, it is certainly adequate to the task of conferring all the properties of life.

Can there be an increment in our biological knowledge in the next fifty years comparable to what we have gained in the past fifty years? Propheteering in the sciences is a precarious enterprise—witness Mr. Bateson—but it is enticing to speculate. We have just begun to understand the molecular mechanisms of heredity. The unanswered questions outnumber by legion those we think we have mastered. How is the apportionment of only half of the DNA into the sex cells achieved? What is the molecular mechanism of dominance? How can one piece of ribbon of DNA gain ascendancy over another almost identical in total composition? Is there a discernible biological impact by structures below the atomic level? And finally, the most imminent question of today's biology: What are the regulatory mechanisms which hold the reins on DNA? Every organism, however complex, contains in every one of its cells all of the genetic information of that particular species. Thus every one of a normal human being's thousands of billions of somatic cells contains the identical number and composition—and consequently, shape—of chromosomes. And yet a cell so endowed, located in the tip of our finger, produces nails, and one in our eye a light-sensitive pigment.

What is the molecular conductor which cues in and out segments of the DNA and thus orders the differentiation of each of the myriads of cells, endowing them with the wondrous variety of attributes essential for life's many tasks, and yet maintains those cells sufficiently akin to be amenable to integration into the harmonious whole that is an individual?

Our present concepts of these mechanisms of regulation are so crude, so lacking in biological reality and therefore so ephemeral it would be wasteful of the reader's time to relate them.

But once these now-hidden mechanisms are revealed and come within the reach of our tools, man may indeed become the master of his corporal fate. The practice of what can only be called molecular surgery might give us new limbs for old, may suppress deleterious genes from our seed, and enhance our creative capacity. Are these perfervid dreams? Perhaps they are to some; but not to those who ask, as Galileo did in 1615, "Who indeed will set bounds to human ingenuity?"

INDEX

INDEX